Statistical Methods for Plant Variety Evaluation

PLANT BREEDING SERIES 3
STATISTICAL METHODS FOR PLANT VARIETY EVALUATION

Edited by R.A. Kempton and P.N. Fox

The present book is the third in the Plant Breeding Series published by Chapman & Hall and follows the very favourable reception that the two earlier volumes have already received. The first book, *Plant Breeding: Principles and prospects,* edited by Hayward, Bosemark and Romagosa, set a wide ranging and authoritative broad view of techniques and approaches being adopted in modern plant breeding. The second, *Selection Methods in Plant Breeding,* by Bos and Caligari, was written to help further secure the scientific basis underlying selection methods as applied in practical plant breeding.

This third book tackles the vitally important area of statistical techniques for data handling and analyses along with, of course, the trial designs needed specifically by the modern plant breeder. The efficiency and effectiveness of plant breeding and selection programmes have often been left unexamined. As a consequence many breeding programmes have made much less impact than they otherwise might have done. Thus, for example, imagined differences form the basis for selection; systematic biases are unwittingly included in the trial designs because the necessary application of statistics has been ignored. Often, because plant breeders in the past have been unwilling to make themselves sufficiently acquainted with the underlying features and philosophy of statistical application (as opposed to theory), they have been unable to at least adopt rudimentary designs for their trials which would allow meaningful analyses. This philosophy has changed greatly over recent times but now, with the widespread availability of computers, thoughts and possibilities are moving very quickly. This book thus aims to provide a timely and well balanced account of statistical methods, data management and handling allied to the potential for effective experimental designs specifically with plant breeding in mind.

As noted in the earlier books the need for plant breeders to be successful has never been greater. The potential future population growth must fill all of us with concern and with a clear indication for the absolute need for plant breeders to succeed. The adoption of new and potentially powerful techniques, such as those provided by biotechnology, plant transformation and molecular biology, are well perceived. But these will only make any real and practical input when applied within the context of more complete approaches to plant breeding. Foremost among the integrated requirements for success is the ability to evaluate any material produced in trials which give clear, informative and applicable results – the need for *Statistical Methods in Plant Variety Evaluation* is obvious.

The present book is aimed, as the first two were, not only at the student who is learning the subject of plant breeding but also the breeder who is trying to actually improve our crop species. It has been written to make clear the exciting possibilities for exploiting modern statistical methods to further the aims central to plant breeding worldwide.

Peter D.S. Caligari
Plant Breeding Series Editor

Statistical Methods for Plant Variety Evaluation

Edited by

R.A. Kempton
Biomathematics and Statistics Scotland

P.N. Fox
CIMMYT
Mexico

Coordinating editor
M. Cerezo, CIHEAM,
Zaragoza, Spain

CIHEAM

CHAPMAN & HALL

London · Weinheim · New York · Tokyo · Melbourne · Madras

Published by Chapman & Hall, 2–6 Boundary Row, London SE1 8HN, UK

Chapman & Hall, 2–6 Boundary Row, London SE1 8HN, UK

Chapman & Hall GmbH, Pappelallee 3, 69469 Weinheim, Germany

Chapman & Hall USA, 115 Fifth Avenue, New York, NY 10003, USA

Chapman & Hall Japan, ITO-Japan, Kyowa Building, 3F, 2-2-1 Hirakawacho, Chiyoda-ku, Tokyo 102, Japan

Chapman & Hall Australia, 102 Dodds Street, South Melbourne, Victoria 3205, Australia

Chapman & Hall India, R. Seshadri, 32 Second Main Road, CIT East, Madras 600 035, India

First edition 1997

Typeset in 9.5/11.5pt Palatino by Columns Design Ltd, Reading
Printed in Great Britain by St Edmundsbury Press, Suffolk

ISBN 0 412 54750 3

A Catalogue record for this book is available from the British Library
Library of Congress Catalog Card Number: 96-70871

♾Printed on permanent acid-free text paper, manufactured in accordance with ANSI/NISO Z39.48–1992 and ANSI/NISO Z39.48–1984 (Permanence of Paper)

Contents

Contributors

J.D. Corbett
Integrated Information Management Laboratory
Blackland Research Center
Texas A & M University System
808 E. Blackland Road
Temple
Texas 76502
USA

J. Crossa
International Maize and Wheat Improvement Center (CIMMYT)
Lisboa 27, Colonia Juárez
Apartado Postal 6–641
06600 México, DF
Mexico

P.N. Fox
International Maize and Wheat Improvement Center (CIMMYT)
Lisboa 27, Colonia Juárez
Apartado Postal 6–641
06600 México, DF
Mexico

A.C. Gleeson
NSW Agriculture
Division of Corporate Services
161 Kite Street
Locked Bag 21
Orange NSW 2800
Australia

R.A. Kempton
Biomathematics & Statistics Scotland
James Clerk Maxwell Building
The King's Buildings
University of Edinburgh
Edinburgh EH9 3JZ
UK

H. Ketata
International Center for Agricultural Research in the Dry Areas (ICARDA)
Germplasm Program
PO Box 5466
Aleppo
Syria

R. Mead
The University of Reading
Department of Applied Statistics
PO Box 240
Earley Gate
Reading RG6 6FN
UK

H.D. Patterson
15 Kings Grove
Longniddry
East Lothian EH32 0QW
UK

P. Portmann
Agriculture Western Australia
Crop Breeding
3 Baron-Hay Court
South Perth
Western Australia 6151
Australia

I. Romagosa
Centre UdL – IRTA
Departamento de Producción Vegetal
Avenida Rovira Roure 177
25006 Lleida
Spain

M. Talbot
Biomathematics & Statistics Scotland
James Clerk Maxwell Building
The King's Buildings
University of Edinburgh
Edinburgh EH9 3JZ
UK

Foreword

The International Centre for Advanced Mediterranean Agronomic Studies (CIHEAM), established in 1962, is an intergovernmental organization of fourteen countries: Albania, Algeria, Egypt, France, Greece, Italy, Lebanon, Malta, Morocco, Portugal, Spain, Tunisia, Turkey and Yugoslavia.

Four institutes (Bari, Italy; Chania, Greece; Montpellier, France and Zaragoza, Spain) provide post-graduate education at the Master of Science level. CIHEAM promotes research networks on Mediterranean agricultural priorities, supports the organization of specialized education in member countries, holds seminars and workshops bringing together technologists and scientists involved in Mediterranean agriculture and regularly produces diverse publications including the series 'Options Méditerranéennes'. Through these activities, CIHEAM promotes North/South dialogue and international co-operation for agricultural development in the Mediterranean region.

Plant breeding is, since 1982, one of the fields in which CIHEAM is conducting a programme of training and promotion of research, through its Mediterranean Agronomic Institute of Zaragoza. The programme includes a one-year advanced specialization course, a series of two-week courses for professionals, and support for collaborative research activities in Mediterranean-relevant species. Many of these activities are organized in collaboration with national and international institutions.

The publication in 1993 of *Plant Breeding. Principles and Prospects*, which initiated the Chapman & Hall's Plant Breeding Series, has become an important feature of CIHEAM's plant breeding programme. We are now pleased to add a further volume to that series: the present book, *Statistical Methods for Plant Variety Evaluation*.

As was the case for the first publication, *Statistical Methods for Plant Variety Evaluation* originates from one of CIHEAM's courses. The course was organized in collaboration with the International Center for Agricultural Research in the Dry Areas (ICARDA) and the International Maize and Wheat Improvement Center (CIMMYT) to deal with the important issue of the improvement of the economic efficiency of plant evaluation trials. The good interaction between a group of lecturers, internationally known for their expertise in this field, and an international group of professionals participating in two editions of the course, has produced excellent results and, as the editors have underlined, has helped to motivate the authors to produce this book.

In their preface, the editors present the objectives and the contents of the book, and describe the essential elements in its creation. This description shows clearly that the book is the result of a combination of contributions and of efforts from individuals and institutions. We would therefore like to thank all those who have directly or indirectly participated in this project. Our special thanks go to all the authors of the chapters for their valuable contributions and for the interest they have shown in this endeavour, and very specially to R.A. Kempton and P.N. Fox, the editors who, with their enthusiasm, dedication and skill, have made this publication possible.

We acknowledge, on this occasion too, the collaboration and interest of Chapman & Hall as well as the Series Editor, Professor Caligari, The University of Reading.

We would like to express our wish that this book will help all those who read it, and especially plant breeders, to plan more efficient experiments and open new opportunities for collaboration.

Miguel Valls
Director
Mediterranean Agronomic Institute of Zaragoza, Spain

Preface

Plant breeding is experiencing major changes in technology and relative economics, the full strategic implications of which are not yet appreciated. But one point is clear: as the costs of generating field, glasshouse and laboratory information continue to rise, the costs of managing that information are falling. Plant breeders must respond by ensuring that the data are collected efficiently using the best experimental practice and trial designs, and are fully utilized using modern methods for their management, exchange and analysis.

The genesis of the book was a course presented in 1991 at the International Centre for Advanced Mediterranean Agronomic Studies (CIHEAM), Zaragoza, Spain, in collaboration with two international agricultural research institutes, the International Maize and Wheat Improvement Center (CIMMYT) and the International Center for Agricultural Research in the Dry Areas (ICARDA). Lecturers came from these institutes, from the Scottish Agricultural Statistics Service (now Biomathematics and Statistics Scotland), the New South Wales Department of Agriculture, Australia, and the Universities of Lleida, Spain, and Reading, UK. The strength of the course lay in the mixing of statisticians, computer specialists and plant breeders – 'tool makers' with 'tool users'. The high level of interest in the course, which was three times over-subscribed, and the geographical diversity of the 30 participants, who came from 19 countries, demonstrated a widespread awareness of the importance of modern statistical and data management techniques. During the computer practical sessions many participants arrived at new biological understanding of their own data. This demonstrated a lasting impact and the course was repeated in 1993.

We thank participants in both courses for their stimulating questions and suggestions. Their interest and enthusiasm helped motivate the lecturers to produce this book. The Director and staff at the CIHEAM training centre, particularly Manuela Cerezo and María Navarro, assisted the participants and lecturers with efficiency, tact and humour.

This book describes modern statistical and trial management techniques for field evaluation of genotypic performance. It focuses on the distinctive goals of plant breeders and the technical problems that they face at each stage of selection. It includes detailed description of field plot techniques and designs for plant breeding trials; of statistical methods for spatial analysis and controlling inter-plot competition; of methods for multi-environment testing and analysis of genotype by environment interaction; and of efficient partitioning of resources among the different stages of selection.

Individual chapters have been edited to provide cross-referencing and eliminate inconsistencies, though the careful reader will notice differences in approach and emphasis which reflect the working backgrounds and concerns of particular authors. Inevitably, a single volume cannot deal with all aspects of plant breeding. In particular, we say nothing about the selection of parents for a crossing programme or the statistical techniques in biometrical genetics, linkage analysis and molecular sequencing which can assist with this choice. Emphasis, moreover, is on selection for yield in annual crops, where the authors' experiences mainly lie, and we do not consider selection for multiple characters in depth or for a single character over several harvests. With these exceptions, however, the book provides a comprehensive treatment of statistical issues which are increasingly important to a successful breeding programme.

R.A. Kempton and P.N. Fox
Zaragoza, Spain

1

Introduction

R.A. KEMPTON AND P.N. FOX

1.1 CHANGE AND CHALLENGES

The last 50 years witnessed a remarkable increase in world agricultural productivity. Nowhere is this more apparent than in the rising yields of the major cereal crops (Fig. 1.1). Indeed, if crop yields today were still at 1970 levels, 250 million additional hectares of land, an area three-quarters the size of India, would be required for world food production. The increase is partly due to changes in cultural practice: improvements in cultivation and harvesting methods due to new machinery, the use of chemicals to control pests, diseases, weeds or crop growth, and the increased use of irrigation and fertilizer. However, a major contribution to higher yield has come from genetic improvements as a result of plant breeding and variety selection. For example, Silvey (1986) estimated that new varieties accounted for approximately half of the increase in yield of winter wheat in England and Wales in the 1960s and 1970s. In the developing world, Baum (1986) estimated that the introduction of new semi-dwarf wheat varieties raised average yields by more than 500 kg/ha (i.e. between 25% and 50%). The International Maize and Wheat Improvement Centre (CIMMYT, 1993) estimates that research on spring bread wheat between 1977 and 1990 contributed US$3 billion in benefits to developing countries through improved yields and disease resistance. Bosemark (1993) concluded that 'better adapted crop varieties are the cheapest, most reliable and environmentally safest way to increase productivity and secure the world's food supply'.

Over the last decade, however, the steady and rapid rise in crop yields appears to be levelling off. This may be partly the result of environmental and economic constraints on the use of inputs: fertilizers, chemicals and irrigation. While in the 1970s world attention focused on industrial pollution, there is now equal concern about the agricultural impact on soil erosion and land degradation, the detrimental effects of agrochemicals and the loss of biodiversity. Past breeding success means that opportunities for exploiting major genetic variation for the main food crops are now more limited. Rapid

Statistical Methods for Plant Variety Evaluation. Edited by R.A. Kempton and P.N. Fox. Published in 1997 by Chapman & Hall, London. ISBN 0 412 54750 3.

Fig. 1.1. Annual world yield of principal cereals 1961–1994 (data from Food and Agriculture Organisation).

genetic advance through molecular techniques seems unlikely, though introgression of alien germplasm may be important in maintaining and increasing genetic variation. Future increase in crop yield is thus most likely to come through the accumulation of small genetic gains and identification of genotypes which can best exploit particular environments. Agricultural environments are themselves changing, as seen, for example, in the increase in mixed farming and the long-cycle graze/grain cereals in rotation with pastures for sustainable agricultural practice.

Since the 'world food crisis' of the 1970s when the Food and Agriculture Organisation (FAO) forecasters threatened widespread famine, the increase in world food production has outpaced population growth and the issue of feeding the world has largely been removed from the political agenda. As a consequence, agricultural research has experienced major cuts in funding. Despite this, future projections of world food production continue to assume improving efficiency and yield, which will need to be research based.

Notwithstanding falls in fertility rates throughout the world, including, for the first time, sub-Saharan Africa, the current world population is predicted to increase by 50% within the next 20 to 25 years. At the same time, world grain consumption in 1995/96 is projected to exceed production for the third year in succession. Grain stocks relative to consumption are now at their lowest level for more than twenty years. Unless the past balance between food production and consumption is maintained, conflicts generated by food and water shortages could make the dominant issues of today – poverty, the environment, population and resource protection – irrelevant.

To meet the challenge of feeding the world in the next decade and beyond, plant breeding teams need to make optimal use of the dwindling resources currently allocated for agricultural research. It is therefore essential to use the best methods for evaluating genotypes in order to maximize genetic advance. These include modern methods for plot and data management, an integrated approach to statistical design and analysis across all stages of the selection programme, and detailed investigation and exploitation of genotype by environment interactions.

1.2 MANAGING THE INFORMATION

The success of plant breeding programmes worldwide increasingly depends on the breeders' ability to handle large populations. At CIMMYT about 6000 crosses are made for bread wheat improvement each year and the resulting genotypes are routinely evaluated in many countries. At a national level, the winter wheat programme at the Plant Breeding Institute, Cambridge, UK, whose varieties in 1980 occupied 70% of the UK acreage, involved approximately 700 crosses, evaluation of 1.4 million spaced plants at F_2 and yield assessment of over 1000 genotypes in trial plots at F_5 (Bingham *et al.*, 1981). Evaluation of such large numbers of plots requires particular attention to efficient field technique and data handling. Trials should be laid out and managed to optimize work efficiency while maintaining small error and bias (see Chapter 2). Machinery for small plots should reflect farm practice as far as possible. Data are best validated at the point of collection so that errors are easily corrected. Computerized data capture devices, incorporating automatic validation procedures, enhance quality control and allow rapid data processing.

Modern data management systems provide the breeder with much more extensive information on which to base decisions (Chapter 3). A database may include information on the pedigree of genotypes and present and past performance in trials, as well as details of the management and environment of individual trials. This information can be used to identify elite parents for crossing, to improve the criteria for the selection of genotypes and finally to identify domains for regional recommendation. Standardization of the way data are organized and stored in database management systems creates opportunities for data sharing as a means of increasing the resources dedicated to crop improvement. For example, genotype × tillage system interactions may be explored over a much wider range using international databases than is possible with more limited national or regional information.

1.3 STATISTICAL DESIGN OF THE SELECTION PROGRAMME

To maximize genetic advance in a crop selection programme, allocation of resources and other issues of statistical design need careful consideration. These issues include: (i) the number and choice of parental crosses; (ii) the number of replicates grown in each trial and the choice of treatment design; (iii) the number and choice of trial locations; and (iv) the length of the selection programme and proportion of genotypes selected at each stage.

Consider, for illustration, the programme used for pedigree selection of winter wheat at the Plant Breeding Institute (Fig. 1.2). (Other breeding systems will differ in detail, but can usually be classified into similar stages from a biometrical point of view.) In the early stages, selection is based largely on negative attributes, as genotypes are discarded on the basis of disease susceptibility, plant height, maturity and other highly heritable traits. The first yield trial occurs in F_5 and is restricted to a single replicate at one location. In F_6 and F_7, with smaller numbers of selected genotypes and more seed available, the number of replicates per trial is increased (usually to three plots of each genotype) and the trials are grown first at four, then at eight, UK locations. Elite material may be evaluated at further locations. After three years

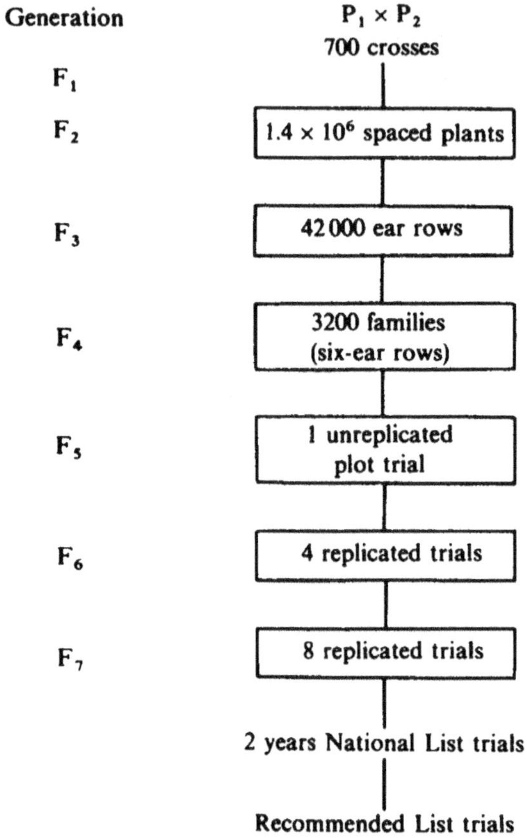

Fig. 1.2. Pedigree selection system for winter wheat followed at the Plant Breeding Institute, Cambridge (from Bingham *et al.*, 1981).

in yield trials, promising genotypes are submitted for up to five years of official trials before recommendation for commercial use in the UK (Patterson and Silvey, 1980).

The resources available at any stage are limited, usually by the number of plots that can be physically managed by the breeding team. The number of plots is a product of the number of genotypes evaluated and the number of plots sown with each genotype. Hence, a breeder may choose to carry out a broad screening of a large number of genotypes, or a more precise evaluation of a smaller number. What is the best strategy? A theoretical investigation provides helpful insight.

For a single trial, the genetic response R to selection (the genetic gain) for a continuous character such as yield can be expressed as

$$R = ih^2\sigma_p \tag{1.1}$$

(Falconer, 1981), where i is the intensity of selection, σ_p is the phenotypic standard deviation of the genotype means in trial, and h^2 is the heritability. The phenotypic variance is the sum of the genetic and environmental variances, $\sigma_p^2 = \sigma_g^2 + \sigma_e^2$, where for a single trial with r replicates and plot error variance σ^2, $\sigma_e^2 = \sigma^2/r$. The heritability, h^2 equals σ_g^2/σ_p^2. The genetic gain may be increased by reducing the environmental variance and thus increasing the heritability. This is achieved by increasing replication. However, since the number of plots is fixed, extra replication will reduce the number of genotypes evaluated and, if the number of genotypes selected is fixed, the selection intensity is reduced through the consequent increase in the selected proportion. Bos (1983a) showed theoretically that if the baseline heritability for a single replicate is high (say, $h^2 > 0.5$) while the selection proportion is not too low (say > 5%), then maximum genetic gain is achieved with only one replicate (see Chapter 6). This result assumes that the genotypes discarded to allow replication represent a random sample of the gene pool with respect to yield, which will often be unrealistically cautious. Furthermore, Bos (1983a) ignored the potentially large additional reduction in environmental variance from using more sophisticated treatment design and analysis in replicated trials. Taking account of these additional factors, we suggest that, when there is sufficient seed, two or three replicates should be used.

The number of trial replicates also needs to be balanced against the number of trial locations. This will depend on the relative sizes of between-location and within-trial environmental variances (Talbot, 1984), and also on the fixed costs involved in each additional location (Sprague and Federer, 1951). Theoretical calculations again suggest restricting replication to two or three plots per trial location.

How well the comparison of genotypes in a breeding programme predicts their potential performance across a region will depend, in part, on the quality of results from individual regional trials. Precise comparison of large numbers of genotypes is difficult because plot heterogeneity across a trial can lead to a large within-trial variance. For replicated trials, this variance can be substantially reduced by dividing the trial area into small artificial blocks and allocating genotypes to blocks in a special way. Some trials may require more complex blocking structures, for example to remove plot variation across both rows and columns, or to investigate genotype interactions with different inputs. Block designs are considered in Chapter 4.

The quality of prediction also depends on the appropriate choice of locations. Suppose that the breeder has identified a domain (e.g. region, silo district or soil type) for which selected genotypes are to be recommended. Trial locations are then chosen to represent the range of environments where the crop is grown within this domain. A random sample, possibly stratified for environmental factors, is ideal but rarely feasible. In practice choice of locations is restricted by accessibility and the need to ensure that trials at remote locations are well managed. Using locations particularly prone to environmental stresses, such as drought or disease, ensures that genotypes are exposed to these stresses at an early stage. This allows the selection programme to be shortened while ensuring that regional recommendations are robust to extreme years.

Yield assessment usually extends over several years or stages, with selections made at each stage. The design question now is how to choose the selection proportions and resources allocated to each stage. Theoretical results (Finney, 1958) indicate that a symmetric scheme with equal proportions of genotypes selected and equal resources allocated at each stage is approximately optimal in the absence of genotype × year interactions. The benefit from using more than three stages is also small in terms of overall genetic gain.

1.4 STATISTICAL MODELS AND PREDICTION

The prediction of genotype yield from trial data requires an underlying statistical model which provides an estimate of the mean yield and the distribution of possible yields.

For a single variety trial, where genotypes are arranged in blocks to control plot heterogeneity, prediction of the genotype yield is usually based on the additive model:

$$\text{Plot yield} = \text{trial mean} + \text{genotype effect} + \text{block effect} + \text{plot error} \quad (1.2)$$

When each block contains a single replicate of each genotype, the predicted yield difference between two genotypes is the difference in their mean yields. However, when the blocks are incomplete, genotypes will not always occur together in the same block, so the means first need to be adjusted for block effects. From the distribution of plot errors across the trial, it is also possible to predict the distribution of difference for each pair of genotypes for different randomizations of genotypes to plots. From this we can estimate the precision of the estimate of mean genotype difference. This precision is improved by good blocking as a significant proportion of the plot error is then removed by block effects. The analysis is described in detail in Chapter 4.

More complex models than Equation 1.2 are sometimes used to reduce the error of prediction and so improve the efficiency of the analysis for individual trials. For example, a covariate such as weed or disease score may be included to provide more information on the variation among plots and a better correction to genotype means (see Section 4.3.4). Plot errors may be described in terms of a spatial model which takes account of correlations between the yields of neighbouring plots (see Chapter 5). In contrast to block analysis which models the fertility trend as a step function (possibly refined using smaller, incomplete blocks), spatial analysis fits a smooth trend to plot yields. Another extension of the simple block + genotype model allows the results to be adjusted for interference which can arise, for example, from competition between tall and short genotypes in adjacent plots (see Chapter 7).

With these models, however, prediction error is not based on randomization and may be underestimated. A complex model may sometimes provide a better fit to plot yields by mimicking the random plot errors as well as the systematic pattern. Comparison of the predicted values with the yields from further replication soon show the improvement in fit to be spurious. Such a model is said to overfit the data and is poor for prediction.

Regional prediction is based on the results from a series of trials which represent the range of environments in which the crop is grown. The simplest additive model is

$$\text{Yield} = \text{overall mean} + \text{genotype effect} + \text{environment effect} + \text{error} \quad (1.3)$$

where the yields are derived from the predicted values for the genotypes in the individual trials or environments. For complete data, where each genotype appears in each trial, the predicted yield difference of two genotypes is given by their mean yield difference, averaged over all trials. For incomplete data, the means must be adjusted to avoid bias, and this becomes more complex when the environments combine different locations and different years (see Chapter 9).

In the additive genotype–environment model, the residual error is in large part due to genotype by environment interaction. More complex models may include non-linear terms which attempt to describe this interaction and so improve prediction. Examples are Finlay–Wilkinson regression and Additive Main effects and Multiplicative Interaction (AMMI) models (see Chapters 8 and 9). While more complex models are more flexible and may give a better description of current data, they often prove to be poorer for prediction. As explained before, this is because the models overfit the data and describe variation which is not characteristic of the wider population of trials for which the prediction is made. This is a particular issue in respect of the AMMI model which has been promoted by Gauch (1992). In our view (Kempton, 1994), the inclusion of multiplicative terms in the model is better reserved for exploring patterns of genotype by environment interaction (see Chapter 8) than used for prediction.

1.5 IMPACT OF GENOTYPE BY ENVIRONMENT INTERACTION

Genotype by environment interaction (GE) is the differential expression of genotypes across environments. It reduces the association between phenotypic and genotypic values, and may cause selections from one environment to perform poorly in another. This forces breeders to examine genotypic adaptation. Chapter 8 describes a number of statistical methods which provide useful tools for illuminating patterns of adaptation and shows how these should be linked to the biology of a particular crop. For example, the different floral structures of wheat and maize explain gross differences in adaptation patterns. Analysis of GE is also used to identify environmental domains within which interactions are small so that recommended genotypes will perform consistently well at all locations. In many cases, however, genotype × location interactions are not consistent over the years, so that recommendation domains are plastic and will change with time.

The genetic basis of GE is still poorly understood, and the evolving capabilities for data management (see Chapter 3) could provide a useful source of genetic and genealogical information. Further progress in analysing GE will also draw on management and environmental data for individual trials, while recommendations will be more finely attuned using data accessed through Geographic Information Systems.

1.6 SUMMARY

This book emphasizes the need for an integrated approach to plant variety evaluation. To produce a successful variety, a breeder must ensure that every stage of the selection process is carried out as effectively as possible. This requires attention to field plot technique, trial design, data management and analysis, and multilocation testing. A shortfall in just one component of the process may lead to the costly introduction of a new variety which shows no improvement over existing varieties in commercial production.

With trial costs escalating, it is also important to maximize the efficiency of the selection process. To do this we need to understand how the different sources of variation contribute to final selection error so that breeding resources can be allocated optimally. This is considered in Chapter 10, which also describes the special selection criteria developed for official variety testing in Europe.

1.7 NOMENCLATURE

The authors have adopted, as far as possible, a standard terminology throughout the book. 'Genotype' is used as the entity under test in the selection programme, in preference to variety, cultivar, line, hybrid or clone, though 'Variety' is also used in Chapter 9 and when it is necessary to distinguish finished varieties in commercial production. 'Checks' are varieties used as standards for genotype comparison. 'Environment' is a general term that covers the conditions under which plants grow and may involve locations, years, management practices, or a combination of these factors. 'Location' is used as a synonym for site or centre. 'Yield' refers to grain yield unless otherwise specified.

2

Field plot technique

P. PORTMANN AND H. KETATA

2.1 INTRODUCTION

The success of a plant breeding programme normally depends on the breeder's ability to handle large populations. This may require compromise between experimental strategies that minimize error in the trial data and operational procedures that allow large quantities of material to be evaluated. Mechanization of plant breeding operations and the increasing availability of computers have had a major impact on the size of populations in breeding programmes. In particular, computers have facilitated much of the routine book keeping. Trial designs, field books, lists for seed preparation and cleaning, field plans and sample labels can all be readily generated by computer and this can greatly assist the planning and management of trials. Most programmes are dependent on such facilities to handle the population sizes necessary to achieve advance.

Experimentation must reflect the farm situation as far as possible, so that results are readily transferable. This principle should always be considered, particularly in developing machinery for experimental use. Fortunately, mechanization for both breeders and farmers is making this ideal achievable.

A common-sense approach which draws on the experience of breeders is required, in conjunction with careful use of resources to maximize the productivity of breeding programmes and minimize the factors that work against success. Creativity in finding better ways to conduct research is a continuing challenge and the life-blood of successful breeding programmes.

2.2 TERMINOLOGY

Every plant breeding centre develops its own terminology to describe the physical components of their trials. The terminology used in this book is explained in Fig. 2.1.

Statistical Methods for Plant Variety Evaluation. Edited by R.A. Kempton and P.N. Fox. Published in 1997 by Chapman & Hall, London. ISBN 0 412 54750 3.

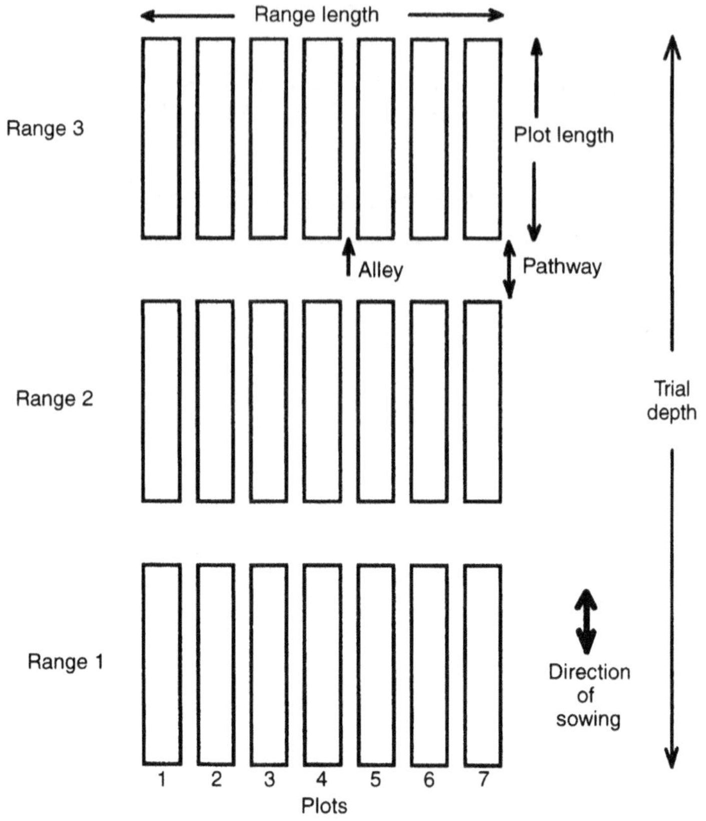

Fig. 2.1. Trial layout.

2.3 WITHIN-LOCATION VARIATION

No two trial locations are identical and rarely is any one trial location uniform across its whole area. Soil variation and the interactions of genotypes with that variation are major factors that breeders must accommodate. One should strive to eliminate or minimize the effect of obvious variation, by the strategic positioning of trials, giving attention to previous history and soil samples. Previous history is often a major source of problems (e.g. roadways, plot versus pathway areas and differing herbicide or fertilizer treatments) and these need to be avoided, as do areas close to trees and turning areas for cultivation equipment. Uniform swards of cover crops or pasture can be used in the cycle after plot trials to minimize such residual effects on soil heterogeneity in subsequent trials.

Otherwise it is a matter of choice and compromise. Restricting the number of genotypes per trial will increase the chances of at least locating complete replicates on a uniform area, but will also increase the fragmentation of data and the problems of comparing all the material under evaluation. Statistical innovations such as alpha designs (Section 4.5.4) and spatial analysis (Chapter 5) have improved the plant breeder's capacity to accommodate soil variation and work with larger numbers of genotypes.

In many cases, genotype by environment interactions are so large that minimization of errors at any one location, although desirable, has only a small effect on the overall picture. Often the breeder will make pragmatic decisions for overall efficiency, rather than for statistical finesse at any one location. There may be more to gain by directing resources towards another location in the programme than reducing the error at any particular location.

2.4 INTER-PLOT INTERFERENCE

When plots are grown side by side, the performance of one plot can affect its immediate neighbour. A common practice, to minimize this problem, is to use unharvested border rows to buffer the harvested plot area from the effects of its neighbour. However, this system has practical disadvantages and may replace one experimental artefact of unknown quantity with another. Often border rows are much more vigorous than the rest of the plot. By their very vigour they can compete with neighbouring rows, which are thus depressed in relation to the mean performance of the plot. This effect is common in cereals, for example, where the yield of individual rows of a plot can oscillate about the mean yield with decreasing amplitude from the outside to the centre of the plot. In general, it is preferable to test material in multirow plots as soon as possible. The more rows in a plot, the less will be the impact of border effects, regardless of whether border rows are harvested. To ease identification in the field and to allow plots to be harvested with minimal cross-contamination, most breeding programmes leave an alley between plots, often equivalent to one missed row. If the alley is too wide, it poses a potential management problem from, for example, weeds or soil erosion.

The need to retain as much seed as possible is often a factor in the ongoing viability of the programme and for this reason it may be most appropriate to harvest the whole plot and ignore the border effects. In practice, variety by border interactions will usually be much smaller than variety by location interactions, and therefore of less importance. The extra seed may allow trials to be sown at one or more extra locations and this may be the better option. Whether or not to ignore border effects must be the choice of the breeder in the light of experience and other needs of the programme.

It may, at times, be strategic to group material to minimize competition. Material may be best grouped into different maturity and height groups or disease reaction types to simplify general field operations and reduce levels of interaction. It may be possible for a breeder to quantify competition effects, but this will require extra field work and would need to be weighed against other options for the use of resources that may give a better overall return.

These options are considered in more detail in Chapter 7.

2.5 THE PLOT

The plot is the basic unit within the trial. It may be many rows wide and many metres long, or it may be a hill plot or a pot in a glasshouse. The size and shape of plot will be influenced by the crop and type of data being collected. Measurements are either made on the whole plot or on a sub-sample of the plot. When sampling within a plot, the size of the sample is dictated by the intra-plot variation and the desired level of precision. Plot size is often determined by a number of other factors such as the amount of seed available, the inherent variation within the field (experimental error) and the character measured. Hill plots can be satisfactory for some characters such as flowering date, plant height, or kernel weight but are less satisfactory for yield and complex quality characters. Plot size will also be determined by the sowing and harvesting equipment. In general, for measuring yield it is desirable to have rectangular rather than square plots. This is particularly so if spatial analysis is used, so that any trends can be maximized in one direction. Often it is the practical requirements that control the size of the plot.

It may be desirable to keep the number of replicates and plot size small when seed supplies are limited, so that each genotype is grown at as many locations as possible. The more locations sampled, the better the breeder can quantify the genotype by environment interactions and effectively select. Single replicate trials (Chapter 6) can increase the number of locations at which material is tested, especially if used with replicated checks to estimate error. However, in such cases, any plots lost through poor experimental procedure will mean that some genotypes are not tested at that location. For more advanced material, more seed is usually available and hence larger plots and more replicates can be used.

2.6 REPLICATION

Many options are available for trial design, some of which involve complex randomization structures. Before using a design, its impact on operational efficiency needs to be considered along with the likelihood of sowing errors and their impact on the final effectiveness of the design. The increased effort or fragility associated with a more sophisticated design may not warrant its use and a simpler design may be more effective.

If the breeder is less concerned about inter-plot interactions than genotype by location interactions and labour or time is limiting, it is worth considering using the same randomization at each location. This can have a major impact on efficiency at seed preparation by allowing trials at all locations to be prepared simultaneously, greatly reducing the handling of seed and need for individual books and seed processing records for each location. It may permit another location to be included. (These benefits will need to be weighed against the possibility of statistical bias in the analysis, see Section 4.10.)

2.7 TRIAL MANAGEMENT

Trials need to be managed both to minimize error and bias and to ensure optimum work efficiency. Often this requires compromise. Except where differential treatments (e.g. rates of nitrogen) are applied, all of the plots should be handled in the same manner. Seed, fertilizers and herbicides should be evenly applied, using machinery where possible. In most cases, applications across plots rather than along them will increase the uniformity of applications. To ensure that research results have practical application, equipment should reflect farm practice. Often this entails the use of specially designed machinery. Considerable ingenuity over the years has resulted in a good range of equipment for most breeding tasks.

For most breeding programmes, increased uniformity has been a secondary, though important, reason for the use of machinery. The major reason has been increased efficiency by saving time and labour. It has been a bonus to discover that machinery has not only improved efficiency but also increased the accuracy of trials and reduced errors from boredom and fatigue. There is much anecdotal evidence from plant breeding programmes world-wide of a reduction of error in trials coinciding with automation.

Above all else, care and precision in the field will have the greatest impact on breeding success. No statistical design or analysis can remedy mistakes in identifying seed at sowing and harvest time. It is important to take the time to mark out trial areas carefully and clearly, both for sowing and subsequent management. Many trials have been damaged or destroyed by subsequent mistakes at sowing, spraying or harvest because of inadequate initial marking of trials. Care at sowing is needed to ensure that plots do not overlap and that seed is not mixed from one plot to the next. This requires regular inspection of equipment during the sowing operations to ensure that there are no blockages or malfunctions.

Most breeding programmes use some type of cone seeder for sowing trials. Care is required to ensure that seed falls evenly around the cone for uniform seed density along the plot. Cones need to be protected from the wind and work best on level ground. If you need to work on slopes, it is necessary to use specially designed cones that dispense the seed simultaneously from two outlets directly opposite each other into the seed distributor, so that bias in distribution caused by the slope cancels out. In this case, the cones complete one half revolution rather than the normal full revolution over the length of the plot.

A common error with breeding trials is plot stagger, particularly where plots are sown in both directions. Stagger is caused by the time delay between seed being released into the cone and it reaching the ground. The faster the seeder is travelling, the greater the amount of stagger. Operators need to calibrate their seeder for the speed of travel and determine how far in front of the start of the plot they need to release the seed. This is best done by releasing seed with the seeding tynes or coulters out of the ground and observing where the seed starts in relation to the start of the plot and adjusting the point of release accordingly.

Sowing depth can have a critical effect on the uniformity of the trial. Soil texture can vary markedly across a trial, greatly affecting depth of sowing. Depth should be checked regularly and adjusted, if necessary, to make sure that it remains relatively constant. Seeding tynes on a seeder can easily bend, creating differences in sowing depth of rows within a plot. The seeder should be regularly checked to make sure that any problems are detected and corrected as soon as possible.

Each task should be planned to maximize efficiency while minimizing human and experimental error. The time of the day and approaching weather are two factors that need to be assessed before starting a trial. Towards the end of the day, it may be wiser to sow or harvest two smaller trials than one large trial if there is a chance of not finishing the larger due to uncertain weather. Spreading the sowing or harvesting of a trial over two days can have a disastrous effect on results. Delay from unexpected rain can result in variations in yield that render the whole trial useless. In extreme situations, operations may be delayed for days or weeks. If a trial must be stopped before it is complete every effort should be made to complete a replicate.

Careful consideration needs to be given to the management of pathways between trials or plots (see Fig. 2.1). Bare areas of soil between trials can cause serious soil variation in subsequent experiments. Pathways should be minimized and trial plots should preferably be sown end on end, or a buffer crop used to over-sow the pathways. End trimming is a major source of variation in small plots. Many breeders use plots two to three metres long. A 10 cm variation about the mean line when trimming the ends of plots can be responsible for up to a 10% variation in yield due solely to differences in plot length; this is larger than the yield advance most breeders expect with a new release.

In Western Australia, plant breeders have built special spray rigs that simultaneously spray both ends of small plots with a 'knock-down' herbicide, just prior to ear emergence, to kill plants outside the harvested plot area. This ensures a constant plot length determined by the boom of the spray rig. In these programmes, constant plot length is considered more important than interactions between genotype and end effects. A mechanical slasher for simultaneously trimming the plots at each end, just prior to harvest, can also be developed to achieve the same end. It is a more cumbersome machine and shifts the labour demand for this task immediately prior to harvest. These are factors that would need to be weighed against the concerns of trimming with herbicide earlier in the growth of the crop. Most breeding programmes have abandoned rotary hoeing pathways for end trimming because of their inaccuracy and the soil conservation problems that are created.

2.8 TRIAL LAYOUT

Sensible restrictions on trial layout can simplify field operations, reduce the chances of error and increase operational efficiency. Where possible, trials should be designed to a standard trial depth to simplify overall management of the field with less risk of errors. A standard trial depth can be achieved by splitting replicates into a number of ranges (see Fig. 2.1) and stacking ranges and replicates behind each other. A 60 m deep trial provides a high degree of flexibility. It allows breeders to use 1, 2, 3, 4, 5, 6, 10, 12, 15, 20, 30 and 60 m plots. It is also a functional length for operating field equipment and accurate driving at sowing. By designing all of the trials to fit into the one standard trial depth, subsequent management of the plots is simplified.

Management is further simplified if the number of different plot lengths is minimized and all of the trials of the same plot length are sown adjacent to one another. This makes for more efficient use of the experimental area as turning space (for equipment used to apply fertilizer and herbicides or end trim plots) usually needs to be allowed between trials of different plot lengths.

With such a structured approach to design and layout, it is possible to eliminate the need for detailed and time-consuming labelling of trials in the field. A good field map is adequate because trials are easily found. The weeks of labour required to prepare labels can be more usefully redirected to productive activities. Seed preparation and sowing similarly become less confusing and more efficient, with considerable reduction in human error.

2.9 DATA

Once the experiment is harvested, data are all that remain for analysis and interpretation. If the data collected are poor then interpretations will be unreliable (garbage in, garbage out). Consideration needs to be given to the most appropriate data to be collected. Often the temptation is to collect too much data and not give appropriate attention to the most important. Filing cabinets and computer files world-wide are filled with data that have been collected because they may be of interest to someone in future. Most is never of interest to anyone and if it is, its existence is unknown to those seeking the information, who will set out to collect the data again, probably in a trial better designed for the purpose (however, see Chapter 3 on initiatives for enhanced data sharing). In general, it is best to collect only the data required to answer the questions posed, when setting up the trial, and plan another trial for other data in the future, if necessary. It is vital to plan the trial carefully from the outset and map out the data required, how and when they are to be measured and collected, and by whom. It is futile to set up an experiment only to find that, when data collection starts, resources to service the commitments are inadequate.

Data may be directly related to the objective of the experiment or may be supplemental data (e.g. rainfall, temperature, soil type, latitude or altitude) that may assist in the interpretation of primary results. Data may be qualitative or descriptive (e.g. dry versus wet, heavy versus moderate disease level) or quantitative (e.g. grain yield kg/ha, tillers/m^2 or protein %). Descriptive data are often invaluable in interpretation of results and care should be taken to observe and record the relevant features and phenomena about the trial locations and conditions. Unwanted effects on the trial (e.g. weed and insect infestations, bird or animal damage) should be carefully noted. The recording of such occurrences should indicate the plots affected, the level of damage, the date of incidence, and the name of the observer. The name allows those interpreting data the opportunity to ask questions for clarification.

Much qualitative data are most usefully expressed in some form of numeric rating. For example, instead of describing something as low, medium or high it is usually better to assign a rating that is readily analysed. Many breeding programmes use a 1–9 scale (often known as the 2–5–8 scale). The main scores are 2, 5 and 8 representing low, medium and high levels. Scores of 1 and 3 allow variation around 2 and similarly 4 and 6 around 5, and 7 and 9 around 8. The recorder needs to keep only three basic scores in mind. Plots that rate a little better than one of the three main categories can be marked up one point and those a little worse down a point from the nearest main category. A zero value can be used if appropriate.

Quantitative data are usually, but not always, better than qualitative data. For example, most breeders require relative flowering dates of the material under test. The traditional practice is to visit the field two to three times a week and note material that has flowered since the last recording. This provides good quantitative data but can result in some 15 or more visits, a major cost on time, especially when the field is distant from the home base. In practice, most breeders do not require the exact day that their material flowers but differences within their material. Western Australian plant breeders, with trials up to 800 km from their home base, have developed a phenology score similar in concept to Zadoks' decimal score (Zadoks, Chang and Konzak, 1974) but designed to be more predictive and linear in relation to time. They rate all their material for development with one visit to each trial location. The score adequately discriminates material with minimal effort. The task is made easier and most relevant by scoring at a location when about 50% of the material is flowering and differences are most readily rated.

2.10 DATA COLLECTION

It is important to record the data using measurement tools of the appropriate accuracy. If the accuracy is too low, significant effects may not be identified, if too high then the data will become unnecessarily complicated (see Chapter 3) and may consume unjustifiable resources in its collection. A breeder needs to apply common sense, using knowledge of the magnitude of variation of the trait, the accuracy of the measuring device and the economic importance of a unit difference.

Data should be checked for any errors made during collection or transcription (Chapter 3). In large programmes it is often not practical to check every data value, but computer programs are available to check for abnormal values and identify out of range data by a number of different tests, e.g. maxima and minima, standard deviations and scatter diagrams. Data should be checked as soon as possible after collection so that data can be remeasured, where possible.

One source of error, transcription of data from field notes into a computer, can be eliminated. There is a range of hand-held computers or data-loggers that can be programmed to record data directly in the field or from electronic balances and other equipment. This not only reduces the frequency of errors but also improves the productivity of the programme. To make such units as easy to use as possible, software for recording data in the field should allow the hand-held computer to simulate the field book, pencil and eraser. Depending on the power and capacity of the computer, it is possible to copy the details of the trial entries and previously collected data onto the computer so that the operator has access to all of the appropriate details of the trials in the field, without carrying a field book. However, most people still find it helpful to produce field books for casual reference.

2.11 DATA ANALYSIS

The type of analysis is primarily determined by the objective of the experiment, the structure of the treatments and the experimental design used. The breeder needs to determine the appropriate design and analysis before commencing the experiment as this may place restrictions or special requirements on how the trials are sown and managed.

When interpreting and summarizing the results of an experiment there are a number of factors that need to be considered.

- Statistics does not solve all of the problems. The most powerful statistical analysis can do little to improve data from badly designed trials or poorly collected data. When used properly in sound experiments, statistics can greatly assist the researcher to avoid subjective judgements.
- Differences between treatments or treatment effects should be subjected to some significance test. The level at which the data is considered significant is a subjective decision of the breeder. Statistical significance does not ensure differences in economic value. This has to be assessed by other means, e.g. economic analysis.
- If the results seem extraordinary, the researcher should try to identify possible reasons before drawing any conclusions. Even if no obvious reason is found, the researcher should point to the unusual results and seek ways of verifying the results by further research.
- The more precise the experiment and the more the experiment samples the environmental range, the more reliable are the results and conclusions. Conclusions drawn from an experiment repeated for three or four years at many different locations in a region are far more robust than those from experiments conducted at one or even a number of locations only once.

Maximizing the efficiency of the operations within a breeding programme is a process of common-sense. Always question whether it is appropriate to continue to carry out activities in the same way as before. Encourage staff to think about easier or more efficient ways of doing the work; often a breeder's staff have a better idea of the work involved and will be more effective at identifying ways of being more efficient. Reassess whether the data being collected is the most appropriate for the end purpose; there may be faster or more appropriate ways of collecting information that will achieve the same end result. Finally, question whether an increase in experimental finesse is worth the effort in the context of the whole evaluation programme carried out over all the locations and years.

3

Data management and validation

P.N. FOX, R. MEAD, M. TALBOT AND J.D. CORBETT

3.1 INTRODUCTION

As the costs of data management and data storage decrease, plant scientists should use advances in computing hardware and software to reap the maximum value from collected information. At the same time, costs of field plots (see Chapter 10) and generating laboratory data generally continue to increase. These changing relationships in costs of research components in crop improvement demand better use of collected data than in the past. Increased communications facilities support data transfer and sharing among scientists. However, unsystematically arranged data cannot be profitably shared.

This chapter outlines the possibilities for systematic data transfer and sharing. Procedures for data validation increase the value of data and the users' confidence in predictions made from them. Well-designed data management systems may strengthen data validation itself, by providing better estimates of expected values than were available previously.

3.2 DATA MANAGEMENT

The advantages of improved data management in plant breeding range from small gains in efficiencies to radically new ways of sharing and employing research findings. A relatively small change is that a breeder now has ready access to information from previous years when making decisions on whether to discard or advance genotypes. In the past, reference to field books from past cycles while, for example, processing current yield trials was difficult because of time and space limitations. Such logistical limitations usually meant that the data were used only once in the selection stream, although greater genetic advance could be expected (Chapter 10) if all of the data generated were considered when selecting genotypes. Now, digital storage and retrieval of data facilitate the aggregation of all relevant data for

Statistical Methods for Plant Variety Evaluation. Edited by R.A. Kempton and P.N. Fox. Published in 1997 by Chapman & Hall, London. ISBN 0 412 54750 3.

making selections. However, as discussed below, careful planning of databases is required.

There are major efficiencies in sharing performance data. International crop databases may remove some of the uncertainty and unpleasant surprises from varietal releases, especially with respect to susceptibility to diseases. For example, all of the susceptible reactions of a genotype world-wide could be reviewed before releasing it.

When the same genotypes are grown in different areas of the world, the yield data can be profitably pooled across international regions, as indicated by the correlated response for irrigated wheat in northwestern Mexico and in the Punjab of the Indian subcontinent (Braun, Pfeiffer and Pollmer, 1992). Such correlated responses may have even greater implications for dryland areas where genotype × year and genotype × location × year interactions are large. The eastern wheatbelt of Western Australia and Settat in Morocco experience large short-term interactions with weather patterns and long-term climatic similarity with each other. Therefore, genotype performance averaged across both locations may reduce the influence of short-term weather fluctuations at one location and thus provide better information on genotype performance for both.

The collective investment in field plot management of germplasm distributed by the International Maize and Wheat Improvement Center (CIMMYT) Wheat Program alone exceeds US$1 million per year. Now, however, information technology for data management is available to put the data generated by such major investments to work in crop improvement.

The same genotypes may be grown in many countries. Expensive genetic tests, particularly related to industrial quality, are still repeated in many different laboratories. There is a huge opportunity for rationalizing costs and sharing information. In addition, in a self-pollinating crop with a low level of outcrossing, such as wheat, many redundant genetic tests can be eliminated through data management. For example, if two parents are monomorphic for a given allele, we can infer that genotypes resulting from crossing these parents also contain the allele. This process, which depends on maintaining a database for known genes in genotypes, can reduce testing to those progeny in which the trait in question is segregating.

Data sharing depends on cooperation within the research community, and scientists need to communicate and cooperate more when organizing their data and making them accessible. Capabilities for doing so are not yet fully appreciated. The increase in computing and networking power, coupled with continually decreasing prices, makes it difficult to define exactly what a data management and validation system will look like in even two to three years. However, it is important to (i) organize existing data; (ii) organize systematic procedures for updating data; and (iii) take advantage of useful hardware and software advances.

Conversion between different scales of measurement is generally possible for across-location summaries provided the scales used to record the data are stored. Confusion between missing values and true zeros is much more serious in data sharing, particularly for disease and lodging scores. We recommend conventions that always record zeros, indicate plots not scored and never leave blank records in a field book.

Crop breeding benefits from germplasm exchange, and relating information to germplasm adds value to such exchange. In the past, data were usually organized on the basis of when, where and by whom they were collected, rather than on the germplasm with which they were associated. Consequently, information from different sources, e.g. national trials, international trials, laboratories and germplasm banks, could not be integrated around identifiable genotypes.

Databases can now link performance in multilocation trials to both environmental and genetic characteristics. Environmental information includes parameters, such as latitude, longitude and altitude, that are invariant at a location across time, as well as factors, such as rainfall and cloud cover, that vary markedly with time and are thus linked in a database to specific crop seasons. The history of crop rotations preceding a trial may also form an important part of the environmental record. Less conventional sources of data originating from satellites include meteorological parameters and infra-red photography, which identifies actively growing crops. Positive dynamic feedback between genetics, phenotypic data, and environmental information will provide new insights into crop adaptation.

The reactions of indicator genotypes, selected either consciously or unwittingly, are important in feedback between genetic and environmental information. For example, if we knew the genetics of boron tolerance and could examine the performance of a few contrasting genotypes for this trait from an extensive performance database (e.g. three decades of accumulated data from the International Spring Wheat Yield Nurseries distributed by CIMMYT), the spatial distribution of boron-related problems could then be examined by comparing the reaction of tolerant and susceptible genotypes. Marked superiority of tolerant genotypes would point to locations where boron toxicity has a large effect. Such inferences could be validated through soil testing at a subset of these locations. Similarly, the performance of known resistant lines relative to susceptible ones across the world would provide bio-assays for pathogens.

The two major types of database structure are relational and object-oriented; the first is described in Section 3.2.1. A brief comparison of the two alternatives is presented here. Most relational databases work on the basis of merging and dividing files or two-way tables, while the object-oriented procedure imposes a more hierarchical structure that is directed to individual data records. The object-oriented structure is particularly appropriate with much larger databases than are commonly used in plant breeding, for example, in satellite weather monitoring, where the continual reforming of tables in a relational structure may be computationally prohibitive. A second generation of relational databases using a foundation of three-way tables also enhances querying capabilities. The principles for manipulating three-way tables are similar to those outlined in Section 3.2.1 for two-way tables.

In the past, links between genetic information from laboratory studies and field data were often unsystematic. The records of thousands of genotypes evaluated in the field were not coupled to the data from more intensive laboratory tests on a few genotypes, principally because of the different scales of operation. Even if field and laboratory researchers were aware of each other's work, the different types of data could not be easily combined in an accessible structure. However, relational databases can cross-reference these two types of information efficiently. Combining old data from traditionally isolated sources can provide plant scientists with new and broader views of their subject. Furthermore, decisions on which data should be stored are now less critical, as data storage and analysis become increasingly cheap relative to data generation.

3.2.1 Databases

A database management system is a computer system for storing, manipulating and retrieving data. In most relational databases, data files consisting of rows and columns are created and stored in a structured manner, which allows examination of inter-relationships among data stored in (physically) separate files. A relational database can manipulate data like a spreadsheet, but the strength of the relational database is its ability to link related entries from many sources. In the past decade, as microcomputers became commonplace, many breeders began to store data in spreadsheets and experienced an immediate improvement in their data management. However, as the amount of stored data increased, the limitations of spreadsheets and the need to use software specifically designed for data management soon became apparent.

The relational approach follows the theory of relational algebra (Ullman, 1982) for the design of the database. A relational database begins with tables made up of columns (containing attributes) and rows (called entities). Each entity has a set of attributes, at least one of which must be a unique identifier of that entity. These unique identifiers are referred to as primary keys. Primary keys are attributes (stored as columns) that appear in different tables to link entities.

Some data redundancy (primary keys included in different tables), implicit in relational models, allows the links between tables. Although repeating columns in separate tables requires space, it is more efficient to store a single key than to repeat identical attributes for every entry. As a simple example, every entry in a yield database for, say, 50 genotypes at a single trial location would have the same location name, latitude, longitude and elevation. A number identifying the trial location repeated in one column could then link the yield table to a table that holds the locational attributes (name, latitude, longitude and elevation) for every trial location.

Two tables without a common key may be linked if each has a foreign key in common with a third table. In seeking data on a genotype with a known common name, CIMMYT's relational model would first access the Wheat Pedigree Management System (WPMS) to obtain unique identifiers for this named genotype. These unique identifiers are called the cross-identification (CID) and selection identification (SID) numbers. SIDs distinguish among sister lines from the same cross. With these identifiers, the nursery, year and entry number for the genotype are found via a second file. The file containing

WPMS identifiers and the nursery data files link through it. Data are then accessed from files using the entry number for the selected year (Fig. 3.1).

The first step in relational database design is the definition of the conceptual model or schema of the relationships between data sets. Care must be taken to ensure that the key variables identify specific sets of attributes. In general, all of the tables contain rows and columns, and every column that is not part of the primary key must be fully dependent on it. For efficiency, the column values should not contain repetitive groups of data. The working rule for most circumstances is that each attribute of a table must represent an observation only about the primary key (Healey, 1991).

Attributes are not normally accessed except via a key variable. If the attributes are accessed without respecting the key variable, a situation can develop where the same primary key accesses different attributes, a violation of set theory. The opposite case, where attributes describe a primary key that does not link to any other file, is perhaps the most difficult situation to repair. These 'islands' of data describing a key are never accessed from linkages in the relational structure and are thus effectively lost. Regardless of size, if a database is not designed correctly, the outcome will be an inefficient database with major flaws.

The relational database structure offers an integrating mechanism to link information generated by the many disciplines involved in agricultural sciences. For example, a specific trial at one location may generate a series of data, including agronomic observations such as days to heading and maturity, disease scores, daily meteorological information, and yield and industrial quality variables. These data need not be stored in a single file. If disease reactions, industrial processing characteristics and pedigrees are in separate tables, they can be linked using nursery, year, location number and genotype identifier as attributes.

Unique genotype identifiers (from common names to crossing and selection histories) allow breeders to exchange information relating to specific genotypes. The quality and quantity of information available to plant breeders are enhanced by combining data from related experiments. Statistical methods may be applied to identify representative and discriminatory environments (see Chapter 8). Such secondary information may increase the value of primary observations, such as disease and yield scores, in a relational database.

Other data sets of importance to plant breeders are generated by germplasm banks and include morphological and agronomic characteristics of accessions. An individual genotype can also be linked to its genetic lineage, including, perhaps, landraces among its progenitors. Such information may assist in identifying new or under-utilized genetic material that would increase genetic diversity. Pedigree or genealogy analysis can help a plant breeder to identify potential sources of materials for improvement.

3.2.2 Geographical Information Systems

A Geographical Information System (GIS) is software for the management of data characterized by their location (e.g. the latitude and longitude of the test

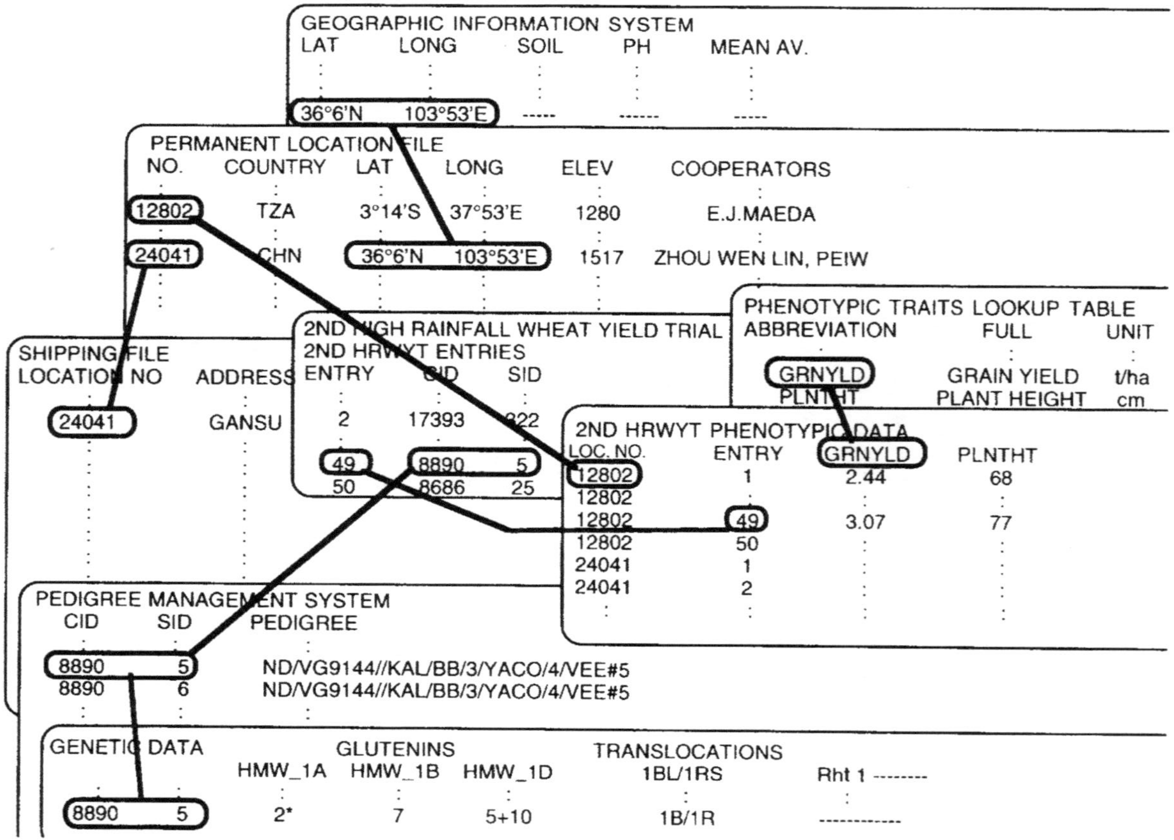

Fig. 3.1. A relational database for wheat, showing common keys linking tables. CID is the cross identifier and SID the selection identifier for a genotype.

location). For plant breeders, the strength of a spatial data management system is its capacity to provide information on test locations that can be used in supporting the analysis of genotype × environment interactions.

For example, temperature, relative humidity, dewpoint and dew duration, and wind direction and velocity all have important and direct influences on wheat rust epidemiology. Recent insights into the relationship between meteorological conditions and disease intensity allow simulation (using long-term meteorological records) of the intensity and frequency of specific diseases or, in some cases, disease vectors (e.g. insect carriers). Mapping the extent and intensity of crop diseases allows characterization of test locations, as well as production areas. Characterization of production areas assists in the setting of research priorities. A GIS can provide temporal (frequency/intensity) and spatial information supporting research efforts.

3.2.3 An example for wheat

A wheat data management system ideally includes the pedigree of each wheat. CIMMYT's WPMS maintains the genetic lineage of each genotype through its CID and SID. This database is then complemented by a series of databases, which include trial data (both replicated and non-replicated), an international cooperator mailing list, data pertaining to industrial processing and molecular genetic studies. Figure 3.1 depicts many of the individual databases accessible through a relational structure. Selection of data from relational databases is made by queries. Queries access the many (physically) separate files through a structured system of keys (the primary and foreign keys mentioned earlier). Using keys, it is possible to move among the various databases and select only the information for a particular need.

With a relational structure, queries can begin with a lead into any of the files. This means that a researcher interested in a geographical area can search for trial locations within that area and then access the trial results and genetic information for the genotypes in those nurseries. Similarly, a pathologist may only be interested in results from trials with more than 500 mm of rainfall during the growing cycle. Such a search is easy using databases built on the relational model.

Links to additional environmental data (typically stored in a GIS) only require the coordinates of latitude and longitude and simplify acquisition of data such as monthly climatic statistics for individual trial locations.

3.3 VALIDATION OF TRIAL DATA

The wider availability and use of trial data in database management systems makes it even more important to ensure that data are properly validated before analysis and storage.

Data collection may involve manual or automatic recording. Errors in manual recording are estimated to occur with a frequency of at least one per hundred numbers, so recording should be automated wherever practicable.

Where data are recorded manually, direct entry to a hand-held or portable computer, rather than a field book, will avoid subsequent transcription errors. However, even with automatic recording, aberrant values will still occur, for example, due to faults in labelling, sowing, plot management or recording equipment, or to external factors such as damage by animals.

Ideally, data validation before the analysis should identify all of the values which ought to be checked for validity. In many cases, it will be possible to substitute a corrected value for an obviously incorrect one. A common error is to record values on the wrong scale, for example, in grams rather than kilograms; this is usually easy to spot and correct. When an automated data validation procedure is used (Section 3.3.6), aberrant values are indicated at the time of recording and so can be immediately checked. However, sometimes a value will appear to be either too large or too small, but the experimenter lacks the confidence to choose a corrected value. In such cases, the unacceptable value should be deleted and treated as missing in the analysis. With modern statistical computing packages, missing values cause no problems.

We first consider ways of identifying aberrant values in individual trials and then look at methods for series of trials.

3.3.1 Common mistakes

Mistakes in data may be identified, both before and after analysis. Systematic checks of data validity should precede or be part of making inferences from the data. However, checks necessarily require subjective judgments, based on knowledge of the variables measured. Validation procedures are best when applied early in data processing and are usually easiest to interpret when dealing with raw, untransformed data. However, in some cases, conversion of raw data (for example, to t/ha) can be revealing.

It helps to understand how numerical mistakes can arise. In manual recording, mistakes often arise from transposing digits, e.g. 413 for 431; repeating digits, e.g. 4413 for 413; or omitting repeated digits, e.g. 413 for 4113. Not all the mistakes have a major impact – 413 for 431 may make little difference to the conclusions and consequently be difficult to spot. However, more serious mistakes can remain undetected if the checking procedures are poor or absent. The following trial yields in kg/ha were detected in a large multilocation trial, after the results were published.

	Block I	Block II	Block III
	.	.	.
	.	.	.
Genotype *18*	3477.08	3380.24	5092.15
Genotype *19*	6448.75	55544.44	7113.86
Genotype *20*	5502.63	4213.26	5718.62
	.	.	.
	.	.	.

The number of recorded digits distracts attention from the important message and probably contributed to the mistake not being identified in this case. Three digits are generally sufficient and six are excessive, given the general inaccuracy acknowledged in measuring yield. The correct value for Genotype *19* in Block II is uncertain, but 5554 would be a sensible correction.

In general, choosing an appropriate correction for a value is a somewhat arbitrary procedure, but in the example above any sensible correction is preferable to retaining the recorded value.

3.3.2 Variables and assumptions

The experimenter should know the likely range of recorded data values for each variable. Values outside these ranges (for example 55544.44) are suspect and should be identified automatically by the data validation procedure. Maximum and minimum values should also be recorded for each data set to see whether the data cover most of the anticipated range. Failure in this respect might raise a question about the range limits or, for a subjective scoring system, indicate that only part of the scale is actually utilized. For example, in reviewing disease data supposedly recorded on a 0 to 9 scale (where 9 represents the high extreme) at a remote location and finding the observed maximum value was 4, the data manager would have to decide whether the disease levels were relatively low at that site, or whether the recorder used a reduced scale, perhaps 0 to 4.

Using computer graphics packages to compare histograms of the current data with historical data on a colour screen can help with these decisions. For example, a researcher could compare the histogram of yield data from a harvest at a location with the histogram for the entire database, or for that location over the years. Other simple checks, such as determining that plant height is greater than ear height in maize, are useful in detecting errors.

In addition to pinpointing rogue values that are not credible in themselves, data validation may also identify values as contextually improbable. For example, consider two sets of results:

(a)	Treatment *1*	587	619	612	576	**466**
	Treatment *2*	597	569	627	702	675
	Treatment *3*	670	609	605	649	583
(b)	Treatment *1*	587	619	612	576	569
	Treatment *2*	597	**466**	627	702	675
	Treatment *3*	670	609	605	649	583

The '466' value is by far the smallest, and we might decide to reject it on general grounds. However, if it is within the acceptable range, it might be acceptable in situation (a), where it occurs for a treatment that gives rather lower values, but not in (b), where, apart from the 466, the other values are markedly larger than those for the other treatments.

The assumptions of the analysis of variance provide a background to contextual assessment of validity. These are that variation of the observed values is similar for all genotypes, that block and genotype effects combine additively, and that the variation of individual values, relative to genotype and block means, has an approximately Normal distribution. In situation (b), Treatment 2 had a high mean value and, because of the low 466 value, a large range or variance. The spread of values for each genotype and any relationship between range and mean value across the different genotypes may indicate rogue values (as well as suggest the possible need for transformation of variables).

The use of residuals from the additive model is considered in the next section.

3.3.3 Residuals

In an initial analysis of variance, there is a second opportunity to check the validity of the data. This involves the residuals, which are the discrepancies between the observed data and those values that could be expected based on the genotype and block means. Formally, the residual for the ith genotype in the jth block is

$$e_{ij} = y_{ij} - \hat{y}_{ij}$$

where $\hat{y}_{ij}$ is the value predicted from the model for data values y_{ij}:

$$E(y_{ij}) = \mu + g_i + b_j$$

For data from complete block experiments, the residuals can be calculated starting from a block × genotype array of data as follows.

1. Calculate the mean for each block.
2. Subtract the block mean from all values in that block, repeating the procedure for each block.
3. Calculate the mean of the values resulting from (2) for each genotype.
4. Subtract the genotype mean from all values for that genotype.

The procedure is displayed in Table 3.1 for the data from an experiment comparing 18 genotypes in four blocks.

For incomplete block designs, the underlying concept is similar, but an iterative procedure calculates the residuals. An example of residuals for an experiment involving six treatments in ten blocks, each block including three treatments, is shown in Table 3.2.

The residuals contain all the information about the variation relative to the model for block and genotype effects. Consequently, examination of the residuals can provide information about the invalidity of individual data values or of the model. This would be shown up by (i) unusually large residuals or (ii) patterns in the residuals, through graphical representation.

However, some characteristics of residuals can make their interpretation difficult. The construction of residuals necessarily means that the sums of the residuals for each block and for each genotype will be zero (Tables 3.1 and 3.2). Hence, if a particular block or genotype includes a large positive residual, it is also likely to include largish negative residuals. Thus in Table 3.2, the

Table 3.1. Calculation of residuals for complete block analysis with block + genotype model

Genotype	*Original data for blocks* (× 10)				*Subtract block means*				*Genotype mean*	*Subtract genotype means to obtain residuals*			
	I	II	III	IV	I	II	III	IV		I	II	III	IV
1	52	53	57	51	+3	+4	+3	−1	+2	+1	+2	+1	−3
2	48	47	51	50	−1	−2	−3	−2	−2	+1	0	−1	0
3	52	51	52	55	+3	+2	−2	+3	+2	+1	0	−4	+1
4	51	51	52	57	+2	+2	−2	+5	+2	0	0	−4	+3
5	45	45	56	52	−4	−4	+2	0	−1	−3	−3	+3	+1
6	48	48	52	52	−1	−1	−2	0	−1	0	0	−1	+1
7	48	51	53	53	−1	+2	−1	+1	0	−1	+2	−1	+1
8	48	46	54	49	−1	−3	0	−3	−2	+1	−1	+2	−1
9	49	52	58	49	0	+3	+4	−3	+1	−1	+2	+3	−4
10	46	41	52	47	−3	−8	−2	−5	−4	+1	−4	+2	−1
11	51	51	53	50	+2	+2	−1	−2	0	+2	+2	−1	−2
12	44	48	50	52	−5	−1	−4	0	−2	−3	+1	−2	+2
13	49	48	51	52	0	−1	−3	0	−1	+1	0	−2	+1
14	50	52	56	55	+1	+3	+2	+3	+2	−1	+1	0	+1
15	52	52	53	54	+3	+3	−1	+2	+2	+1	+1	−3	0
16	49	49	55	51	0	0	+1	−1	0	0	0	+1	−1
17	52	51	61	56	+3	+2	+7	+4	+4	−1	−2	+3	0
18	48	49	53	44	−1	0	−1	−8	−2	+1	+2	+1	−6
Mean	49	49	54	52									

Table 3.2 Residuals from block + genotype models for an experiment with incomplete block design

Block	*Treatment*					
	1	*2*	*3*	*4*	*5*	*6*
I	−4.2	+0.4	+3.8			
II	−1.5	−0.3		+1.9		
III	+0.5		0.0		−0.5	
IV	+2.9			−3.0		0.0
V	+2.3				−2.6	+0.4
VI		+1.9	−1.6			−0.3
VII		−2.8		+2.6	+0.2	
VIII		+0.8			0.0	−0.8
IX			−0.1	−2.9	+3.0	
X			−2.1	+1.4		+0.7

largest residual is −4.2 for Block I with Treatment *1* and the second largest is +3.8 also for Block I with Treatment *3*. This interdependence of residuals can make detecting individually extreme values and patterns difficult. In Table 3.1, most residuals lie in the range ±3 and there is no excessively large value; nevertheless, it would be worth checking on the combination of Genotype *18* in Block IV to see if there are grounds for suspecting the original value of 4.4.

Plotting residuals against the predicted value (based on block and genotype means) may show trends, for example, a greater scatter of residuals associated with larger means (or small means). A trend would cast doubt, not on individual values, but on the assumption (see Chapter 9) of homogeneous variance for all genotypes. This might suggest some data transformation before analysis. Plotting the residuals against the predicted values for the data of Table 3.1 did not show any trend.

Computer packages usually have facilities for highlighting extreme residuals and plotting residuals to detect patterns. Less usual are facilities for looking at residuals in relation to the physical arrangement of experimental plots. However, examination of residuals plotted in field order is often informative, both from the point of view of detecting rogue values and for identifying groupings of positive (or negative) residuals that might indicate block patterns (Chapter 4) or spatial patterns in plot productivity (Chapter 5). In the same way that GIS techniques are used for mapping on a global scale, using coordinates of latitude and longitude, they may also be applied for graphically depicting trends in productivity (or residual values) within trials using row and column numbers as Cartesian coordinates.

The residuals from Table 3.1 are displayed in plot order in Table 3.3 and show clear patterns. In Block III, there is a division between negative residuals in the left half and positive residuals in the right half. There are clumps of positive residuals at the left hand end of Block I and (possibly) in the middle of Block IV. Clumps of negative residuals occur on the left of centre in Block II and at both ends of Block IV. Note also that the residual −6 at the right hand end of Block IV seems less extreme relative to the next two plots in

Table 3.3. Genotype codes (in italics) and residuals in field plot order for the data in Table 3.1

Block I	*11*	*3*	*8*	*10*	*13*	*17*	*16*	*5*	*14*	*18*	*2*	*6*	*4*	*7*	*12*	*9*	*1*	*15*
	+2	+1	+1	+1	+1	−1	0	−3	−1	+1	0	0	+1	−1	−3	−1	+1	+1
Block II	*15*	*2*	*11*	*17*	*5*	*12*	*10*	*9*	*4*	*13*	*1*	*14*	*7*	*8*	*16*	*6*	*3*	*18*
	+1	0	+2	−2	−3	+1	−4	+2	+2	0	0	+1	+2	−1	0	0	0	+2
Block III	*12*	*15*	*3*	*2*	*16*	*7*	*13*	*11*	*1*	*6*	*18*	*8*	*5*	*9*	*10*	*17*	*14*	*4*
	−2	−3	−4	−4	+1	−1	−2	−1	−1	−1	+1	+2	+3	+3	+2	+3	0	+1
Block IV	*10*	*1*	*16*	*9*	*17*	*7*	*3*	*8*	*2*	*12*	*13*	*15*	*6*	*5*	*14*	*11*	*4*	*18*
	−1	0	−1	−4	0	+1	+1	−1	+3	+2	+1	0	+1	+1	+1	−2	−3	−6

Block IV. These patterns suggest that smaller blocks should have been used and, in fact, these data come from an experiment where each replicate block was split into smaller blocks (design shown in Chapter 4, Table 4.12).

Another example of plot residuals in field position is shown in Table 3.4. The design is an incomplete block alpha design using blocks of five plots (Chapter 4, Table 4.11). The residuals shown have been modified to include block effects and show patchiness. This representation also enabled the detection of a previously unsuspected rogue value, the residual +11 in the 12th row and 2nd column.

Table 3.4. Residuals in two-dimensional field layout for an incomplete block design

Replicate 1		*Replicate 2*		*Replicate 3*	
3	−1	−8	−6	−3	−2
3	0	−6	−7	−2	−6
0	2	−5	−5	1	0
2	−1	−6	−2	−1	0
2	4	−2	−2	4	3
8	2	−4	−2	−4	−1
3	0	−4	−3	−5	1
5	1	−1	0	0	−2
−1	−2	−8	−4	−2	−4
2	−3	−5	−1	−6	0
0	1	−10	−5	−1	−2
1	11	−7	−6	−5	−2
−1	4	−2	−5	−3	2
1	4	−10	−4	0	−1
3	5	−7	−1	−3	5
5	7	−3	1	0	3
−3	3	1	2	1	1
2	3	−1	1	−5	2
2	6	6	3	−3	2
1	6	−2	1	−1	−1
4	4	4	1	−4	−1
1	2	1	3	−2	1
3	7	4	8	−1	1
3	7	5	4	0	3
4	4	4	6	1	0

3.3.4 Validation with several variables

It is often easier to identify possible rogue values by considering the values of two related variables jointly. An effective technique is to plot each pair of variables, either for the original data or for the residuals after an initial analysis. There are two advantages from such a plot.

- There will often be some pattern for the joint variation of the two variables and it is then easy to see a discrepant observation (pair of values) relative to the two-dimensional pattern.

- If the values for a particular plot are extreme for both variables, then looking at both together may provide some assurance of consistency.

In Fig. 3.2, the distribution for a pair of variables is shown in a two-dimensional plot and the one-dimensional distributions are beside the axes. In each of the single variable distributions, the observation F is clearly the smallest but in neither one-dimensional display would it be thought to be unusually extreme. In the joint, two-dimensional plot, point F looks much more separate from the rest of the distribution. More dramatically, observation C is clearly discrepant and needs investigation.

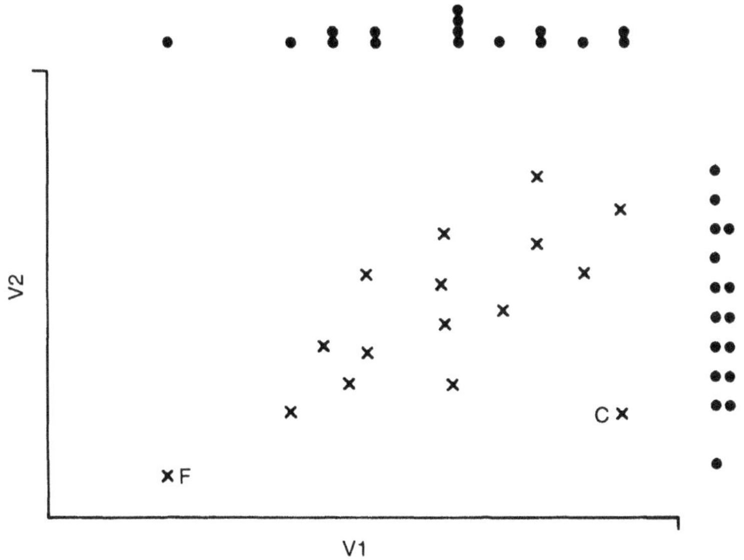

Fig. 3.2. Scatter plot and marginal plots for variables V1 and V2, with two outliers, F and C, identified.

3.3.5 Validating data from many trials

The main objective in assessing the quality of data from trials is to ensure that the results provide a reasonable basis both for predicting relative genotype performance in the future and for studying GE interactions. In practice, most trials fall prey to one hazard or another which makes them unsuitable in some degree. However, trials are costly (Chapter 10) and it is important to extract as much of the information as possible. The pertinent questions are how to recognize trials or data that are atypical and how to deal with them. Here, procedures are described for routine monitoring of genotype × environment (GE) tables and for deciding when to delete aberrant results.

The coefficient of variation (CV), i.e. the plot standard deviation, divided by the mean and multiplied by 100, is often used for assessing the quality of data from a trial. However, the CV on its own can be a poor indicator of quality. Trials that have been under disease, pest or other pressures and, as a consequence, provide important genotype discrimination, often produce high CVs. Also, high CVs tend to be associated with low yielding trials and the use of CVs as a criterion for rejecting trials may discard a higher proportion of these trials, leading to bias in estimates.

A number of alternatives to the CV can be applied to the screening of GE tables. A multivariate approach explored later in the section treats environments as separate variates and calculates principal components to check for association with other factors. Another approach, which can be used with tables that include missing values, involves the analysis of GE residuals. The first stage of this procedure takes the GE table and for each element in the table removes the corresponding genotype and environment effect. The calculation is analogous to forming residuals from the genotypes × blocks array in Tables 3.1 and 3.2.

A residual measures the extent to which a genotype has performed better (+) or worse (−) than expected at a particular location. As with individual trials, the residuals are calculated on the assumption that the effects of environment and genotype are additive rather than, for example, multiplicative; this may not always be appropriate when, say, yields in one environment are much higher or lower than in other environments. It is important that residuals are correctly interpreted. Thus a large positive residual indicates that, relative to its performance elsewhere, a genotype responds especially well in a particular environment, not that the genotype is necessarily the best for that environment.

Dividing a residual by the standard deviation (SD) of all residuals and comparing this with values from a t-table provides an approximate check on the relative significance of a residual. For GE tables with 50 or more cells, a rough rule-of-thumb is that t-values of 2.0, 2.7 and 3.5 are significant at the 5%, 1% and 0.1% levels, respectively. This check is only approximate because the residual being tested contributes to the SD and so the numerator and denominator involved in the calculation of the t-value are not independent. However, with 50 or more values in a table, the validity of the check should not be seriously compromised.

While a residual tells a story about a single genotype, the collection of residuals associated with an environment can tell something about the environment as a whole. For example, a wide range of residuals might indicate unusual performance in an environment. Several statistics can summarize the information represented by the residuals for an environment. The standard deviation ratio is the SD of the residuals for an environment relative to the SD of the residuals for the rest of the table. The average SD ratio is 1.0; an environment with above average variation will be indicated by a ratio much greater than 1.0; an SD ratio of much less than 1.0 is usually associated with an environment where yields are low and differences between genotypes are small.

An indication of the expected range of the SD ratio comes from examining F-values in statistical tables, since the SD ratio can be viewed as the square root of an F-value. For example, with a GE table having ten genotypes and four environments, the degrees of freedom for the F-value for genotypes would be 10 − 1 = 9 for the numerator and (10 − 1) × (4 − 2) = 18 for the denominator, an extra degree of freedom having been deducted for the environment which is examined. The appropriate *F*-value at the 1% significance level is 3.6, which gives an SD ratio equivalent value of $\sqrt{3.6} = 1.9$. In the case of tables with four or more environments, the SD ratio criterion depends principally on the number of genotypes, as follows:

No. of genotypes	10	20	40	60	100
SD ratio criterion	1.9	1.6	1.4	1.3	1.2

Another statistic for checking trial results is the coefficient of correlation between the genotype yields in an environment and the genotype means over other environments; this statistic can be used to assess how closely the yields in an environment follow average yields from all other environments. Correlations of $r = 0.50$ or more are to be expected from yield trials from environments of broadly similar background. The statistical significance of departures from $r = 0.50$ depends on the number of genotypes:

No. of genotypes	30	40	60	100
Min. acceptable r ($P \leq 0.01$)	0.05	0.13	0.21	0.28

In addition to the CV, other statistics for individual trials can contribute towards the validation of GE tables. The genotype *F*-ratio measures how effective a trial has been in discriminating among genotypes. The *F*-statistic is calculated as the among-genotypes mean square in the analysis of variance for that trial, expressed as a ratio of the residual term. A small *F*-ratio suggests that genotypic performance has not been consistent among replicates within locations. This inconsistency contributes to the error term and small, non-significant *F*-ratios are often associated with high CVs and plot-to-plot variation, but may also indicate mistakes in field operations. A further statistic presented as part of the validation procedure is the environment mean. A high CV can be discounted to some extent if it occurs in a trial with very low yields.

Some of the statistics calculated for environments can also be produced for genotypes. Thus the SD ratio for a genotype can serve as a useful measure of genotype stability in the performance across environments.

The application of these check statistics to the validation of GE tables is illustrated in Table 3.5. For ease of interpretation, the residuals in the body of the table are expressed as a percentage of the grand mean of the table. Thus a residual of nine indicates that the genotype yielded 9% above its expected performance from all environments and relative to all genotypes. The SD of the residuals is 6.4% and an approximate criterion for judging the significance of the values in the table at the 5% probability level is calculated as 2.0 × 6.4 = 12.8%. Residuals equal to, or greater than, 12.8% in absolute terms indicate unusual performance of the genotype in that environment.

The summary statistics at the bottom of Table 3.5 need to be interpreted together. For example, a significant SD ratio may be associated with a large CV, which suggests that the atypical features of genotype performance in the

Table 3.5 Residual analysis (as % of mean yield) for a GE table

Genotype	*SD ratio*	*Trial number*					
		10	*11*	*12*	*14*	*16*	*18*
Gerbel	0.9	−1	−3	5	−6	9	−4
Igri	0.9	9	−3	−3	4	−6	−1
Halcyon	0.8	−9	2	5	−2	1	3
Panda	0.9	−3	−0	−5	5	−6	9
Topaz	1.5	−9	3	4	−15[a]	10	2
Emerald	0.4	1	−1	−2	2	4	−4
MMG8048/61	1.6	−13[a]	14[a]	−2	10	−6	−0
CWB131/5/1	0.7	2	−8	1	0	0	4
CWB312/81/3	1.2	8	0	−11	4	−8	5
NFC267/85	1.0	10	0	5	−6	−4	−3
NFC49/83/5	0.6	−3	5	1	2	−6	1
51504/2W	0.6	4	−2	−6	1	2	0
Lady	0.7	1	−7	−1	7	−0	−1
RG5458	1.1	3	5	−4	−9	10	−4
FDE7854	1.0	−1	1	8	4	−5	−9
Gaulois	1.3	−4	−2	3	−15[a]	9	3
SD ratio		1.1	0.9	0.8	1.3	1.1	0.7
Correlation r (× 100)		57	34	60	29	76	62
Genotype F-ratio[b]		NS	**	**	NS	**	**
CV (%)		9.1	4.7	4.4	10.5	5.0	3.9
Environment mean (t/ha)		6.3	6.9	6.2	5.7	8.8	7.4
Correlation (× 100) with genotype scores:							
Lodging		−44	1	49	−80	80	15
Mildew		−2	−54	−8	−6	31	−7
PCA loadings (axis 1)		0.1	0.0	−0.5	0.6	−0.5	0.2

[a] Residuals that are atypical when measured against a t-value at the 5% probability level.
[b] NS, not significant; ** significant at 1% level.

environment is due to replicate-to-replicate variation rather than broader GE factors. On the other hand, a large SD ratio coupled with a low correlation coefficient and a non-significant genotype F-ratio, as for Trial 14, requires closer examination.

A decision on whether to exclude a data item, or the results for a complete trial, from summary reports should rarely, if ever, be taken on statistical grounds alone. The main criteria for ignoring results should be agronomic, with statistical procedures acting as pointers to possible problems. The first aim should be to save as much information as possible, if necessary by omitting results for some genotypes rather than a whole trial. The following procedure for dealing with aberrant values is recommended. First, the reasons for the problem should be sought, perhaps from internal trial records where these are available. Otherwise, it may be possible to examine the relationship between, for example, yield residuals and other genotype characteristics, e.g. disease resistance, based on information from other sources. As an illustration, in Table 3.5 the residuals for Trial 14 are negatively correlated with genotype lodging resistance scores, suggesting that the relatively high SD ratio may be a consequence of lodging damage in the trial.

If the cause of the aberrant values is expected to occur in agricultural practice, then the results are best left to stand. However, if the aberrants are by-products of the testing procedure, e.g. bird damage on plots of selected genotypes, that would not have the same influence in a field crop, then the affected results should be removed or adjusted, for example, by covariance analysis. Where there is no obvious explanation for the discrepancies and the trial has an acceptable CV, it is safer to retain the information, at least for the estimation of genotype means. Provided that there are at least four environments, a single aberrant environment will not usually have a major effect on the ranking of genotypes in a trial series. However, an aberrant environment can have a major influence on GE interactions, especially when these are analysed by multivariate methods. In principal components analysis (PCA, see Chapter 8), for example, an aberrant environment can dominate the first or second principal component. In such cases, it is good practice to carry out the multivariate analysis both with, and without, the aberrant environment.

Multivariate methods are useful tools for detecting aberrants in data sets where multiplicative or more complex interactions overlie the additive genotype and environment effects. Such situations occur with GE tables that include a number of very low-yielding environments. Fitting appropriate models can help to reveal the few environments with a dominant contribution to GE interactions. It can also highlight and justify an individual outlying data value. For example, a biplot (Chapter 8) from PCA will highlight a large specific interaction of a genotype in one environment. The associated data may appear aberrant under the additive model, but may fit comfortably within a multiplicative framework.

PCA represents one of the more helpful multivariate techniques for detecting aberrants, and there are several ways of exploiting the information provided by PCA. For example, Table 3.5 gives the loadings for the first principal component (PCA1) for environments: the extreme value of 0.6 for Trial 14 indicates its unique contribution to GE variation. Another method, useful in identifying values which may cause distortions in interpretation, is to plot PCA1 against genotype and environment mean yields. Influential values are likely to appear isolated from the main body of points in the plot.

To summarize, statistical methods can highlight aberrant data in GE tables, but a decision to exclude data should rest chiefly on whether factors causing the outlying values might also occur in agricultural practice. If in doubt, it is safer to err on the side of including suspect data than to risk discarding valuable information: 'suspect' environments have often subsequently proved to be crucial in selecting genotypes with special properties, once the underlying biological reasons for the atypical responses are determined.

3.3.6 Automating data validation

Data are often generated automatically, with observations transferred directly from the measuring device in computer-readable form or entered directly to a hand-held computer in the field. The use of computers as data capture devices allows data to be automatically validated as they are recorded, when errors are most easily rectified and unusual observations can be explored further (Dickson and Talbot, 1986).

Automated validation systems are likely to have the greatest value when some or all of the following circumstances apply: (i) long sequences of data are recorded; (ii) the volume or complex structure of the data does not allow the operator to identify discordant values easily by eye; (iii) the operator is present to investigate any unusual values; (iv) the discordant values have to be identified immediately if corrective action is to be taken; (v) the consequences of error are non-trivial.

Automated data validation should bring together the experience of recorders and data analysts. The validation procedures should be dynamic, providing the capacity to develop and learn from the previous data recorded and from the operator's responses. In effect, the automated procedures should serve as an expert system: if an observation is initially detected as aberrant but is subsequently justified by the recorder, the automated system should be able to infer that similar observations are also acceptable.

To validate individual values statistically as they are recorded, a prediction function is required to indicate the value to be expected, and a credibility function to assess how unusual is the observation. Both functions should be based on experience with past data.

The prediction function estimates the expected value of a new observation and will depend on the structure of the data that are recorded. For data from a designed experiment with treatment factors, a predictor of an observation might be the mean of all previous observations with the same treatment. When no such observations are available, the predictor might be the mean of all previous observations. A prediction function for observations recorded in time might be a weighted linear regression of past observations, with weights decreasing into the past, to take account of trends in the data over time.

A credibility function estimates the extent to which the recorded value may depart from the expected value without being considered unusual. The distribution of departures of observed from predicted values will often take the Normal form, but the system should allow other distributions. The credibility function is not a test for outliers, but measures the surprisingness of an observation. It should allow for improvement in knowledge about variation as recording progresses, using short-term experience of the current data set, and longer-term experience of other similar data sets. For example, a variance estimate which is a weighted function of both long- and short-term variation might be used, so that for the $(i + 1)$th observation:

$$\text{VAR} = \text{VARH} + f(i)\,[\text{VARC}(i) - \text{VARH}]$$

where VARH is the long-term variance from historical data sets; VARC(i) is the variance for the current data set after i observations; and $f(i) = i/c$ for $i < c$ and $f(i) = 1$ for $i \geq c$. The constant c is set to the minimum number of observations required to provide a reliable estimate of variation for the current data set (when VAR = VARC).

When several variables are recorded on each unit, it is desirable to look at all variables jointly to assess their credibility (see Section 3.3.4). In some circumstances, it may be possible to apply tests to identify specific sources of error. For example, the moisture content of a crop may be estimated by weighing a fresh sample, then drying and reweighing the sample. If several samples are taken, the dried weights may be expected to vary linearly with the fresh weights. If two replicated samples for the same treatment give very different dry matter percentages, then the sample which shows the largest deviation from the dry weight/fresh weight relationship across treatments is more likely to be aberrant.

A tacit assumption in data analysis is that the recorded data are representative of a background population and validation aims to provide the assurance that this is so. Until recently, the data analyst could do no more than act as a detective, identifying the flaws and assembling the evidence that something was wrong, but not correcting identified errors. One of the benefits of automated data collection is that statistical expertise can now be used positively in the correction of errors.

4

Design of plant breeding trials

R. MEAD

4.1 INTRODUCTION

In this chapter we examine the important ideas used in designing field plot trials to compare genotypes. The plots used for such a trial will be relatively small, including usually several rows of the crop, though sometimes a single row or even a single plant may be used as the plot. The number of genotypes may be between ten and several hundred. In many trials two,.identically managed, replicates will be sufficient; at other times three or four such replicates may be appropriate. In later stages of the selection programme, there may be two replicates at each of several sites, or additional replicates under different management systems at a single site.

We restrict our attention to replicated field plot trials – unreplicated trials are considered in Chapter 6. We start with the general concepts and principles, discussing the principles in the context of genotype yield trials (Sections 4.2 and 4.3). Sections 4.4 to 4.8 deal with the control of plot-to-plot variation by blocking, the most efficient methods of designing blocked experiments and the analysis of the resulting data. Finally, in Sections 4.9 to 4.11, we briefly consider some ancillary aspects of design for genotype trials.

4.2 EXPERIMENTAL DESIGN CONCEPTS AND MODEL ASSUMPTIONS

4.2.1 Concepts

The important concepts for the statistical design of experiments are (i) replication; (ii) control of variation between plots; (iii) randomization; (iv) factorial treatment structure; (v) choice of levels for quantitative factors. The first three concepts are important in genotype trials. Additional factors, representing controlled variation in the environment in which the selected varieties will be grown, may occasionally be included in the trials, but the general

Statistical Methods for Plant Variety Evaluation. Edited by R.A. Kempton and P.N. Fox. Published in 1997 by Chapman & Hall, London. ISBN 0 412 54750 3.

concept of factorial structures has less relevance to plant breeding experiments. All five concepts are covered extensively in other books, such as Mead (1988).

4.2.2 Model assumptions

The primary purpose of a plant breeding trial is to assess differences among genotypes for particular traits (e.g. yield). We assume that experimental plots have similarities which enable us to group them in homogeneous blocks, and this leads to an additive model:

$$\text{Plot yield} = \text{Trial mean} + \text{Block effect} + \text{Genotype effect} + \text{Plot error} \quad (4.1)$$

The assumption of additivity implies that both the differences between blocks and the differences between genotypes remain consistent over a trial. In particular:

- The difference between the yields for two genotypes should be similar for each block in which the two genotypes occur together.
- If genotypes A and B occur in block I, A and C in block II and B and C in block III we can estimate the difference between genotypes B and C in two ways. The $B - C$ difference in block III should be similar to the difference between $B - A$ in block I and $C - A$ in block II.
- If the genotypes in two plots, one plot in block I, the other in block II, had been exchanged then the total yield from the two plots would have been the same.

A second assumption which simplifies the principles and practice of the analysis of blocked experiments is that the error terms in Equation 4.1, which represent the random variation between plots, are homogeneous in size and unaffected by the particular genotype. In other words, although the genotypes may differ in average yields, the yield variation between the plots should be similar for all of the genotypes in the trial.

The standard method of analysis is based on least squares estimation for general linear models. This method is a component of most statistics package and provides estimates of yield differences among genotypes along with standard errors of the differences. These estimates have the properties that (i) they are unbiased; (ii) their standard errors are estimated from the residual mean square in the analysis of variance; (iii) the overall significance of the variation of yields between genotypes may be assessed from the variance ratio of the mean square for genotypes (allowing for block differences) relative to the residual mean square; and (iv) an unbiased estimate of the random variance, σ^2, of plot errors is given by the residual mean square. The mean square for the genotypes in (iii) is calculated after adjustment for block differences. In the analysis of variance for general block-genotype designs, the sums of squares for block and genotype effects will take different values depending on the order in which block effects and genotype effects are added to the model.

4.3 PLOTS, REPLICATION, RANDOMIZATION, COVARIANCE

4.3.1 Plots

Plot size and shape will often be dictated by agricultural practice and machinery (Chapter 2). Statistical considerations suggest that the use of many, smaller plots is preferable to fewer, larger plots. This statistical preference assumes a given total area for experimentation and recording. If plots are made too small the results may be unrealistic (see Chapter 7), but plot size should be as small as is practically sensible.

4.3.2 Replication

The number of plots (r) for each genotype should be chosen to provide an acceptable level of precision for comparisons between genotypes. The replication, r, will be largely determined by strategic considerations of the use of resources (see Chapter 10). However, it is important to consider whether the planned level of replication may be expected to give standard errors which are acceptably small.

The standard error (SED) for comparing two genotype means, without complications introduced by blocking, is

$$\text{SED} = \sqrt{(2\sigma^2 / r)}$$

where σ^2 is the random plot-to-plot variance. To make the SED acceptably small we need an approximate value for σ^2, and some idea of how small 'acceptably small' should be. There will usually be information from previous trials on the likely size of σ^2. When a trial is planned to detect a difference of at least d between two treatments, the usual rule (Mead, Curnow and Hasted, 1993) is to choose r to make the SED no bigger than $d/3$. In genotype trials, formal tests of significance have little, if any, relevance. We know there are differences between genotypes, and we select a proportion of the genotypes for further testing regardless of the size of the differences. However, for effective selection, the SED for a difference between two genotype means should still be small relative to the difference (D) between the true yields for the best genotype and a standard or median genotype. A reasonable target could be an SED no bigger than $D/6$.

4.3.3 Randomization

The random allocation of treatments to plots is a necessary requirement for the valid estimation of precision in an experiment. This is particularly important when the calculation and interpretation of $F-$ and $t-$ statistics are a major part of the analysis but it is a sensible procedure for any experiment, and should be part of the protocol for a genotype trial. Detailed randomization procedures for the practical designs recommended in Section 4.5 are considered after discussion of these designs.

4.3.4 Covariance analysis

In many experiments, useful additional information about patterns of plot-to-plot variation will become available during the course of the experiment. This information may be used to improve the precision of treatment differences by covariance analysis.

The additional information will usually become available either as a discernible spatial trend or by identifying a set of unusual plots in the trial. Plots at one side of an experimental area may differ from the main body of plots; some plots may be affected by pests to a greater extent than others; some patchiness of the experimental area may become apparent as the crop grows. Whatever the forms of additional information, it is essential that apparent differences between plots be recorded. The recording may be quite crude, such as a three – or five – point score of damage or reduced apparent fertility. Even a two-value covariate (good/bad or damage/no damage) may be extremely useful (see Chapter 3).

An example of a simple covariance adjustment is shown in the following experiment to compare yields of eight strawberry genotypes (Table 4.1). The trial occupied four ranges of eight plots and genotypes were arranged in a randomized complete block design with blocks corresponding to ranges. Plots at the right-hand side of each range were lower yielding than the rest. When this pattern was investigated it was found that a hedge bordered the right-hand end of the experiment. Using the reciprocal of the distance from the hedge as a covariate improved the precision of the experiment, the standard error of a difference between two genotype means being reduced from 1.01 to 0.52. There were also some changes to the ordering of, and differences between, the genotypic means (Table 4.2). Note that, in this case, the covariance model assumed that the hedge suppressed yields below the level expected for the main part of the field, so the adjusted yields are generally higher than their unadjusted values.

Table 4.1. Experimental plan and yields (kg/plot) for eight strawberry genotypes (labelled *E, F, G, M, P, R, S, V*) in four ranges of eight plots

Range	*Plot*							
	1	*2*	*3*	*4*	*5*	*6*	*7*	*8*
1	*G*5.8	*V*6.3	*R*4.9	*F*6.5	*S*4.5	*M*5.2	*E*6.5	*P*3.8
2	*E*6.9	*P*7.6	*M*7.9	*S*5.6	*G*7.0	*V*5.5	*F*4.0	*R*2.7
3	*V*7.6	*F*6.4	*R*5.0	*G*6.9	*P*7.4	*E*5.3	*S*5.2	*M*3.2
4	*E*7.5	*S*7.0	*M*6.1	*P*7.2	*G*6.5	*F*5.6	*V*5.8	*R*1.4

Table 4.2. The eight genotypes from Table 4.1 with means in descending order of yield (kg/plot) before and after adjustment for plot position

Without adjustment		*With adjustment*	
E	6.55	*P*	7.64
G	6.55	*E*	7.16
P	6.50	*V*	6.93
V	6.30	*G*	6.89
F	5.62	*M*	6.82
M	5.60	*F*	6.37
S	5.58	*S*	6.20
R	3.50	*R*	5.42
SED	1.01	SED	0.52

4.4 BLOCKING – CONTROL AND ADJUSTMENT

The primary reason for using blocks in designing experiments is to control the effects of plot-to-plot variation. If we fail to recognize and use the patterns in the variation then the experiment will provide worse information in two ways, as described below.

Consider a simple experiment for comparing five genotypes using 20 plots and assume that 10 plots are good (high fertility) and 10 poor (low fertility). First suppose that we recognize the pattern of good and bad plots in advance, define four blocks, two containing good plots and two bad plots, and use a complete block design [Table 4.3(a)]. Each genotype occurs in each block and is therefore grown on two good and two bad plots. The comparison between genotype means is 'fair' and the analysis of variance provides an estimate of error variation, s^2, from which block and treatment differences are eliminated.

What happens if we do not use blocks? If we are unlucky some genotypes will be allocated mainly to good plots and others mainly to bad plots [Table 4.3(b)]. This will result in an overestimation of the yield for some genotypes (*E* and *D*) and underestimation of others (*C* and *A*). Alternatively we could be extremely lucky and obtain, through random allocation, a 'fair' design [Table 4.3(c)], in which comparisons between genotypic means are not affected by the good/bad plot differences. However, each genotype will show large variation between the plot yields and consequently the error mean square in the analysis of variance (with no block sum of squares eliminated) will be larger. The failure to recognize and use blocks will therefore give larger standard errors for the genotype comparisons than those achieved with good blocking.

The three outcomes from the three designs of Table 4.3 are summarized in Table 4.4. In practice, failure to use blocks will usually result in a mixture of the two disadvantages of (i) distortion of relative means and (ii) inflated standard errors.

Table 4.3. Designs with and without blocking

	Genotypes				
	A	*B*	*C*	*D*	*E*
(a) With good and bad blocks					
Block I (good)	good	good	good	good	good
Block II (good)	good	good	good	good	good
Block III (bad)	bad	bad	bad	bad	bad
Block IV (bad)	bad	bad	bad	bad	bad
(b) No blocking – unlucky allocation					
	good	bad	bad	good	good
	bad	good	bad	good	good
	bad	good	bad	bad	good
	bad	bad	bad	good	good
(c) No blocking – lucky allocation					
	good	good	bad	good	bad
	bad	good	good	bad	bad
	bad	bad	bad	bad	good
	good	bad	good	good	good

Table 4.4. Benefits of blocking

Design	*Differences between genotype means*	*Standard errors of differences*
Using blocks	Unbiased	Small
No blocks; unlucky	Biased	Somewhat larger
No blocks; lucky	Unbiased	Much larger

The effective control of variation through blocking usually requires fairly small block sizes, while most selection trials use large numbers of genotypes. This leads almost inevitably to the use of incomplete block designs. Comparisons between genotype means will then require an adjustment to allow for differences between blocks.

4.4.1 Adjustment for block differences

The method of least squares for obtaining estimates of genotype differences from an incomplete block design is described in, for example, Mead (1988) and analysis routines are available in many statistical packages. We illustrate the underlying concept for an example with observations on six genotypes in two blocks of four plots:

Block I: *A* 17.1; *B* 21.4; *C* 18.6; *D* 21.9
Block II: *B* 19.2; *C* 17.4; *E* 20.3; *F* 20.8

Overall the means for the two blocks are similar (block I 19.75; block II 19.48), and it would be easy to ignore any differences between blocks and calculate mean values for the six genotypes:

A 17.1; *B* 20.3; *C* 18.0; *D* 21.9; *E* 20.3; *F* 20.8

showing *D* to have the highest yield followed by *F*. However, the yields for the two genotypes occurring in both blocks are clearly higher in block I than in block II. The differences (block I – block II) are 2.2 for genotype *B* and 1.2 for genotype *C*, giving an average block difference of 1.7. This suggests that block I has an advantage of 0.85 (half 1.7) and block II a disadvantage of 0.85. For a valid comparison of genotypes we adjust block I yields down by 0.85 and block II yields up by 0.85 giving

A 16.25; *B* 20.30; *C* 18.00; *D* 21.05; *E* 21.15; *F* 21.65

This alters the relative assessments of the genotypes, with both *F* and *E* now apparently better than *D*.

Block adjustments are very similar to covariance adjustments for simple situations. For many blocks and many genotypes, the philosophy of adjustment is the same as in this simple example, but computation will be greater.

4.4.2 Indirect comparisons

When genotypes in different blocks are to be compared, an equivalent approach to block adjustment is the concept of indirect comparisons. In Section 4.4.1 we are able to compare genotypes *D* and *E* by comparing each with *B* (or *C*). Thus in block I $(D - B)$ is 0.5, while in block II $(E - B)$ is 1.1. Hence we could estimate $(E - D)$ as $1.1 - 0.5 = 0.6$. We could also compare *D* and *E* indirectly through genotype *C* or, for greater efficiency, use both *B* and *C* to make the indirect comparison:

Block I		Block II	
D	21.9	E	20.3
$(B + C)/2$	20.0	$(B + C)/2$	18.3
Difference	1.9	Difference	2.0

Estimated $(E - D)$ difference $2.0 - 1.9 = 0.1$

We can also calculate the standard error of this estimate of the difference between *E* and *D*. The variance of the difference $D - (B + C)/2$ is

$$\frac{\sigma^2}{1} + \frac{\sigma^2}{2} = \frac{3\sigma^2}{2}$$

where σ^2 is the variance between the plots within a block. The variance of the difference $E - (B + C)/2$ is the same and the two differences are independent. Therefore the variance of the estimate of $(E - D)$ is

$$2\left[\frac{3\sigma^2}{2}\right] = 3\sigma^2$$

We can now measure the efficiency of the indirect comparison procedure. For a pair of genotypes in the same block (e.g. *E* and *F*), the variance would be $2\sigma^2$. Our indirect comparison variance is $3\sigma^2$, so the indirect comparison gives two-thirds of the information available from direct comparison.

What about the comparison of *B* with *D*? We have a direct comparison in block I and there is further indirect information from comparing *D* in block I with *B* in block II. The least squares estimate of $(D - B)$ (equivalent to the block-adjusted difference) can be derived algebraically from the least squares equations and the variance of the estimate is $7\sigma^2/4$.

We now compare variances of estimates of genotype differences for this design example with those from an unblocked experiment with the same number of plots per genotype (Table 4.5). If blocking is successful, the error (plot to plot) variance for unblocked design, σ_0^2, will be larger than the variance σ^2 for the blocked design. When $\sigma^2 < \sigma_0^2$ the $(B - C)$, $(A - D)$ and $(E - F)$ comparisons will be better in the blocked design; in addition the $(A - B)$ type of comparison will be better in the blocked design if $7\sigma^2 < 6\sigma_0^2$; the $(A - E)$ type of comparison will be better if $3\sigma^2 < 2\sigma_0^2$. The critical factors of 6/7 and 2/3 are defined to be the design efficiencies for the blocked design for the particular comparisons. We discuss design efficiency and precision again in Section 4.6.

Table 4.5. Variances of genotype differences for seven genotypes *A* to *F*, in blocked and unblocked designs with respective error variances σ^2 and σ_0^2

Comparisons	*Blocked*	*Unblocked*
$B-C$	$2\sigma^2/2$	$2\sigma_0^2/2$
$A-B, A-C, D-B, D-C,$ $E-B, E-C, F-B, F-C$	$7\sigma^2/4$	$3\sigma_0^2/2$
$A-D, E-F$	$2\sigma^2$	$2\sigma_0^2$
$A-E, A-F, D-E, D-F$	$3\sigma^2$	$2\sigma_0^2$

4.5 THE CONSTRUCTION OF INCOMPLETE BLOCK DESIGNS

In this section we show through practical examples that constructing block designs can be simple. We start with designs for small numbers of genotypes to illustrate the basic principles and later consider larger numbers.

4.5.1 Equalizing direct comparisons

An important characteristic of incomplete block designs is the number of direct comparisons (within blocks) for each pair of genotypes. Ideally, there should be equal numbers of direct comparisons for all pairs of genotypes.

Suppose we wish to compare six genotypes in six blocks of four plots. Each genotype should occur four times. The steps for constructing the design in Table 4.6 are as follows.

1. Genotype *1* is allocated to the first four blocks (any four blocks could be used).
2. The 12 remaining spaces in these blocks are to be divided between the five remaining genotypes (as equally as possible).
3. Choose (arbitrarily) genotypes *2* and *3* to occur three times with genotype *1* and genotypes *4*, *5* and *6* twice with genotype *1*.
4. Allocate genotype *2* to blocks I, II, III and V, and genotype *3* in blocks I, II, IV and VI.
5. If genotypes *4* and *5* are allocated to blocks I and II then genotype *6* must go in blocks III and IV with genotypes *4* and *5* (to satisfy point 3), and allocations to blocks V and VI then follow automatically.

Table 4.6. Optimal design for comparing six genotypes in six blocks of four plots

Block					
I	*II*	*III*	*IV*	*V*	*VI*
1	*1*	*1*	*1*	*2*	*3*
2	*2*	*2*	*3*	*4*	*4*
3	*3*	*4*	*5*	*5*	*5*
4	*5*	*6*	*6*	*6*	*6*

The design is even easier to construct by choosing which pairs of genotypes to leave out of successive blocks, i.e. (*5,6*), (*4,6*), (*3,5*), (*2,4*), (*1,3*) and (*1,2*). These six genotype pairs have three direct comparisons and the other genotype pairs two, which is as close as possible to equality.

For a second example, consider seven genotypes in seven blocks of four plots. This time, after allocating genotype *1* to the first four blocks, there are 12 spaces remaining in those blocks to be divided among the six remaining genotypes. The ideal of constructing a design, in which each genotype pair has exactly two direct comparisons, appears possible. It is achieved in Table 4.7 through the following steps.

1. Allocate genotype *2* to two blocks with genotype *1* (I and II) and to two other blocks (V and VI).
2. Allocate genotype *3* to block I and to three further blocks, one of which must include genotype *1* and one genotype *2*: the blocks for genotype *3* are blocks III (or IV), V (or VI) and VII.
3. Allocate genotype *4* to block I; the three further blocks for genotype *4* must include each of the first three genotypes exactly once (IV, VI and VII).
4. If genotype *5* goes in II and III it must also go in VI and VII.
5. Genotype *6* then goes in II and IV and consequently in V and VII.
6. Finally genotype *7* fills the last four places in blocks III, IV, V and VI.

Table 4.7. Balanced design for comparing seven genotypes in seven blocks of four plots

Block I	II	III	IV	V	VI	VII
1	*1*	*1*	*1*	*2*	*2*	*3*
2	*2*	*3*	*4*	*3*	*4*	*4*
3	*5*	*5*	*6*	*6*	*5*	*5*
4	*6*	*7*	*7*	*7*	*7*	*6*

In Table 4.7 all of the genotype pairs have exactly two direct comparisons. A design with equal numbers of all direct comparisons is called a balanced incomplete block (BIB) design and should be used whenever available. Unfortunately, suitable BIB designs rarely occur for genotype trials.

Classical BIB designs are defined in terms of the number of genotypes, g, the block size, k, the treatment replication, r, and the number of direct comparison for each treatment pair, λ. Balance occurs when λ is the same for all pairs of treatments. These definitions imply

$$\lambda(g-1) = (k-1)r$$

so it is uncommon for combinations of g, k and r to produce an integer value of λ. When the number of genotypes in a trial is large, the target of equalizing the direct comparisons for all genotype pairs remains important, though rarely achievable exactly.

4.5.2 Resolvable designs

For ease of management, a genotype trial will usually be designed in complete replicate blocks, each replicate being split into a number of incomplete blocks. Designs with small, incomplete, blocks within each complete replicate block are called resolvable designs and, although they are usually less precise than unrestricted incomplete block designs, the loss of efficiency is small and generally outweighed by the greater management convenience.

For resolvable designs, a second practical design concept relates to the division of genotypes into subsets for the incomplete blocks within each replicate: the subset of genotypes in each incomplete block in one replicate should be evenly spread among blocks in each other replicate.

Consider an example to compare 12 genotypes in three replicates with each replicate split into two blocks of six plots. An agronomist who had been taught the practical concept quickly produced the design in Table 4.8, which almost satisfies the requirements for equal spread. Genotypes *1, 2, 7, 8, 9* and *11* in block I in replicate 1 are split evenly between blocks III (*1, 8* and *11*) and IV (*2, 7, 9*) in replicate 2 and also between blocks V (*1, 7* and *9*) and VI (*2, 8, 11*) in replicate 3. However, block III in replicate 2 is not evenly split between blocks V (*1* and *6*) and VI (*3, 8, 11, 12*) and the equal spread concept is not fully achieved. A single change, swapping genotypes (e.g. *3* and *4*) between blocks V and

VI, produces the required design (though we will see in Section 4.6.1 that the agronomist's original design is almost as efficient).

Table 4.8. Nearly optimal design for comparing 12 genotypes in six blocks of six plots

Replicate 1		*Replicate 2*		*Replicate 3*	
Block I	*Block II*	*Block III*	*Block IV*	*Block V*	*Block VI*
1	3	1	2	1	2
2	4	3	4	4	3
7	5	6	5	6	5
8	6	8	7	7	8
9	10	11	9	9	11
11	12	12	10	10	12

4.5.3 Lattice designs

One particular class of resolvable incomplete block designs has been used in genotype trials for many years. Lattice designs, first proposed by Yates (1936), are restricted to trials where the number of genotypes, g, is the square of an integer and the appropriate block size is $\sqrt{g}$. The basic property of the lattice design is that in any two replicates the division of the g genotypes into blocks in one replicate is such that each block includes exactly one genotype from each block of the other replicate.

Table 4.9 shows the first three replicates of a lattice design for 25 genotypes. The construction of the first two replicates is trivial. In the second replicate, block VI consists of the first genotype in each block of the first replicate; block VII includes the second genotype in each block, and so on. For the third replicate we select genotypes from each block of the first replicate, moving one place along in each successive block. Thus block XI contains the first genotype of block I, the second of block II and so on.

If six replicates are constructed to satisfy the basic lattice property (equal spread) then a balanced design in which each genotype occurs exactly once with each other genotype is obtained. A balanced lattice design is the most efficient design for the particular combination of block size and number of genotypes, but in most practical circumstances we are restricted to fewer replicates than required for balance. In such cases, the incomplete lattice with two, three or four replicates is still the ideal design.

When the number of genotypes to be assessed is not a perfect square, some extra genotypes must be added or existing genotypes deleted if a lattice design is to be used. However, the recent development of more flexible alpha designs makes this inefficient practice unnecessary.

4.5.4 Alpha designs

The development of alpha designs by Patterson and Williams (1976) removes the restrictions on the number of genotypes and its relation with the

Table 4.9. Three replicates of a lattice design for comparing 25 genotypes in blocks of five plots

Replicate 1					
Block I	1	2	3	4	5
Block II	6	7	8	9	10
Block III	11	12	13	14	15
Block IV	16	17	18	19	20
Block V	21	22	23	24	25
Replicate 2					
Block VI	1	6	11	16	21
Block VII	2	7	12	17	22
Block VIII	3	8	13	18	23
Block IX	4	9	14	19	24
Block X	5	10	15	20	25
Replicate 3					
Block XI	1	7	13	19	25
Block XII	2	8	14	20	21
Block XIII	3	9	15	16	22
Block XIV	4	10	11	17	23
Block XV	5	6	12	18	24

block size required for lattice designs. To demonstrate the principle of alpha design construction, we first consider the situation where the number of genotypes (g) is a multiple of the block size (k), $g = sk$. The genotypes are split into k sets of s genotypes per set. For 28 genotypes in blocks of four plots, $s = 7$ and we write down the four sets of seven genotypes:

Set 1	Set 2	Set 3	Set 4
1	8	15	22
2	9	16	23
3	10	17	24
4	11	18	25
5	12	19	26
6	13	20	27
7	14	21	28

Then, as for a lattice design, we construct blocks in successive replicates so that (i) each block includes one genotype from each set; and (ii) for each replicate the genotypes in each block have not occurred together in a block in any previous replicate.

For the first replicate, we choose the seven blocks to be, I the first genotype of each set; II the second genotype of each set; ... ; VII the seventh genotype of each set. For the second and subsequent replicates we rotate the lists of genotypes in the several sets so that the sets occur in different positions relative to each other. For example, the second replicate might be based on the rearrangement:

Set 1	Set 2	Set 3	Set 4
1	*9*	*17*	*26*
2	*10*	*18*	*27*
3	*11*	*19*	*28*
4	*12*	*20*	*22*
5	*13*	*21*	*23*
6	*14*	*15*	*24*
7	*8*	*16*	*25*

Set 1 is unchanged, set 2 is moved up one position, set 3 is moved up two positions and set 4 is moved up four positions. The blocks for the second replicate are formed from the first, second, ... , seventh positions in the cycled sets.

For the third replicate, set 2 is moved up three places from its original position, set 3 six places and set 4 five places:

Set 1	Set 2	Set 3	Set 4
1	*11*	*21*	*27*
2	*12*	*15*	*28*
3	*13*	*16*	*22*
4	*14*	*17*	*23*
5	*8*	*18*	*24*
6	*9*	*19*	*25*
7	*10*	*20*	*26*

The numbers of places by which each set is rotated are chosen so that each pair of sets occurs in different relative positions in each replicate (e.g. for sets 3 and 4, in the first replicate genotype *15* occurs with genotype *22*, in the second replicate with genotype *24*, and in the third with genotype *28*). The resulting three replicate design is shown in Table 4.10.

Alpha designs are most effective when the block size, k, is less than the square root of the number of genotypes, g, and hence less than s. When $k > s$, some genotype pairs occur together in a block in more than one replicate, which can lead to a greater range of standard errors for the genotype differences.

The particular choice of rotations for the sets for each replicate is important. Some rotation patterns will provide more efficient designs than others. Basic generating arrays for producing the rotation patterns for designs with up to 100 genotypes are given by Patterson, Williams and Hunter (1978) for optimal alpha designs. A computer program, ALPHA+ (Williams and Talbot, 1993), is also available to provide optimal designs for up to 500 genotypes. An example of a design produced by ALPHA+ for 50 genotypes in three replicates using blocks of five plots is shown in the pattern of the actual trial layout in Table 4.11.

If the number of genotypes, g, is not an exact multiple of the block size k, a modified method of construction is used. First we construct a design for the smallest value g', which is greater than the number of genotypes

Table 4.10. Alpha design for comparing 28 genotypes in 21 blocks of four plots

Replicate 1				
Block I	1	8	15	22
Block II	2	9	16	23
Block III	3	10	17	24
Block IV	4	11	18	25
Block V	5	12	19	26
Block VI	6	13	20	27
Block VII	7	14	21	28
Replicate 2				
Block VIII	1	9	17	26
Block IX	2	10	18	27
Block X	3	11	19	28
Block XI	4	12	20	22
Block XII	5	13	21	23
Block XIII	6	14	15	24
Block XIV	7	8	16	25
Replicate 3				
Block XV	1	11	21	27
Block XVI	2	12	15	28
Block XVII	3	13	16	22
Block XVIII	4	14	17	23
Block XIX	5	8	18	24
Block XX	6	9	19	25
Block XXI	7	10	20	26

and which is also an exact multiple of k ($s = g'/k$). The surplus (dummy) genotypes are all included in the last set of s genotypes. This ensures that these unwanted genotypes will never occur together in a block in the constructed design. For the actual design we simply omit the dummy genotypes, so that in each replicate there is a mixture of blocks of sizes k and $(k - 1)$.

Suppose, for example, that we want a design with 18 genotypes using four replicates with blocks of size 4 (or smaller). We set up the initial array of four sets of five genotypes, with three further arrays with rotations (0, 1, 4, 2), (0, 2, 3, 4) and (0, 3, 2, 1):

$$\begin{bmatrix} 1 & 6 & 11 & 16 \\ 2 & 7 & 12 & 17 \\ 3 & 8 & 13 & 18 \\ 4 & 9 & 14 & 19 \\ 5 & 10 & 15 & 20 \end{bmatrix} \begin{bmatrix} 1 & 7 & 15 & 18 \\ 2 & 8 & 11 & 19 \\ 3 & 9 & 12 & 20 \\ 4 & 10 & 13 & 16 \\ 5 & 6 & 14 & 17 \end{bmatrix} \begin{bmatrix} 1 & 8 & 14 & 20 \\ 2 & 9 & 15 & 16 \\ 3 & 10 & 11 & 17 \\ 4 & 6 & 12 & 18 \\ 5 & 7 & 13 & 19 \end{bmatrix} \begin{bmatrix} 1 & 9 & 13 & 17 \\ 2 & 10 & 14 & 18 \\ 3 & 6 & 15 & 19 \\ 4 & 7 & 11 & 20 \\ 5 & 8 & 12 & 16 \end{bmatrix}$$

The resulting design after omitting genotypes *19* and *20* is shown in Table 4.12(a) in an unrandomized form.

Table 4.11. Alpha design for 50 genotypes in three replicates, each replicate consisting of ten blocks of five plots. The trial was laid out as six ranges of 25 plots (each plot containing three rows of plants)

Plots	*Replicate 1*		*Replicate 2*		*Replicate 3*	
	Block					
	I	II	XI	XII	XXI	XXII
1	*45*	*22*	*21*	*5*	*25*	*31*
2	*34*	*39*	*2*	*11*	*48*	*50*
3	*8*	*26*	*19*	*41*	*35*	*3*
4	*17*	*31*	*16*	*40*	*9*	*42*
5	*42*	*4*	*31*	*3*	*32*	*14*
	III	IV	XIII	XIV	XXIII	XXIV
6	*5*	*36*	*7*	*35*	*26*	*17*
7	*28*	*24*	*46*	*17*	*49*	*12*
8	*25*	*44*	*39*	*38*	*41*	*16*
9	*14*	*38*	*25*	*13*	*23*	*19*
10	*29*	*3*	*45*	*23*	*33*	*46*
	V	VI	XV	XVI	XXV	XXVI
11	*12*	*27*	*20*	*12*	*13*	*19*
12	*43*	*50*	*32*	*26*	*6*	*18*
13	*33*	*23*	*18*	*6*	*7*	*8*
14	*48*	*15*	*50*	*14*	*15*	*36*
15	*21*	*46*	*34*	*1*	*37*	*29*
	VII	VIII	XVII	XVIII	XXVII	XXVIII
16	*10*	*7*	*4*	*36*	*22*	*2*
17	*32*	*30*	*8*	*47*	*28*	*47*
18	*37*	*41*	*44*	*15*	*40*	*43*
19	*11*	*9*	*33*	*48*	*34*	*45*
20	*47*	*16*	*37*	*22*	*21*	*24*
	IX	X	XIX	XX	XXIX	XXX
21	*49*	*2*	*49*	*42*	*5*	*39*
22	*1*	*18*	*28*	*43*	*38*	*27*
23	*19*	*6*	*10*	*27*	*30*	*44*
24	*20*	*40*	*24*	*29*	*4*	*11*
25	*13*	*35*	*9*	*30*	*20*	*1*

4.5.5 Randomization

In incomplete block designs there are three stages of randomization: (i) random allocation of genotypes to genotype numbers; (ii) random allocation of block sets of genotypes to actual blocks within replicates; (iii) random allocation of genotypes to plots within each block. Stage (iii) is familiar to experimenters who have used complete block designs. Having decided the subset of genotypes to occur in a block, we need some method of allocating genotypes to plots: a formal random allocation procedure should be used. Stage

Table 4.12. Alpha design for 18 genotypes in four replications using blocks of three or four plots, (a) unrandomized, (b) randomized and laid out with each replicate occupying a range

(a) Before randomization

		Replicate 1		Replicate 2		Replicate 3		Replicate 4
Block	I	*1 6 11 16*	VI	*1 7 15 18*	XI	*1 8 14*	XVI	*1 9 13 17*
	II	*2 7 12 17*	VII	*2 8 11*	XII	*2 9 15 16*	XVII	*2 10 14 18*
	III	*3 8 13 18*	VIII	*3 9 12*	XIII	*3 10 11 17*	XVIII	*3 6 15*
	IV	*4 9 14*	IX	*4 10 13 16*	XIV	*4 6 12 18*	XIX	*4 7 11*
	V	*5 10 15*	X	*5 6 14 17*	XV	*5 7 13*	XX	*5 8 12 16*

(b) In field layout after randomization within blocks and within replicates

6 1 16 11	*12 17 7 2*	*10 15 5*	*18 3 13 8*	*9 4 14*
14 5 6 17	*2 8 11*	*9 3 12*	*4 10 13 16*	*7 18 1 15*
8 14 1	*5 7 13*	*12 6 4 18*	*15 16 2 9*	*11 17 10 3*
11 4 7	*9 17 13 1*	*16 5 8 12*	*14 18 2 10*	*6 3 15*

(ii) is specific to incomplete block designs. With s subsets of genotypes and s blocks of k plots in each replicate, we require a system for allocating subsets to blocks: again a random allocation procedure provides this. Stage (i) is not essential for a single experiment. Because, in most cases, the chosen design will not be balanced, some genotype comparisons will be (a little) more precise than others. We may therefore choose deliberately which genotypes should be better compared. If we do not have clear preferences we allocate genotypes to genotype numbers randomly. Semi-random allocation is also possible; we might have several control genotypes and might accept that the precision of comparisons between the controls should be less good, in order to improve the comparisons of the other genotypes with the controls. The randomization procedures in ALPHA+ allow such restrictions of randomization.

The randomization procedure can be illustrated for the design of Table 4.12. In the first replicate, the genotype subsets I(*1, 6, 11, 16*), II(*2, 7, 12, 17*), III(*3, 8, 13, 18*), IV(*4, 9, 14*) and V(*5, 10, 15*) are randomly allocated to the five blocks of plots, the new ordering being I, II, V, III, IV. Then the genotypes are randomly allocated in each block, e.g. $(1, 6, 11, 16) \rightarrow (6, 1, 16, 11)$. Both random allocation procedures are applied for each replicate, to produce the results in Table 4.12(b). This design was previously discussed in Chapter 3 and shown in Table 3.3. The set of genotypes for the trial includes five controls which were allocated to different blocks to give poorer control–control comparisons and better control–other comparisons. Comparing Table 3.1 with Table 4.12(b) we see that genotypes *1* to *5* in Table 4.12(b) are replaced in Table 3.3 by (*3, 5, 4, 1, 2*). The randomization of the other 13 genotypes produces

$$(6, 7, 8, 9, 10, 11, 12, 13, 14, 15, 16, 17, 18)$$
$$\rightarrow (11, 16, 12, 9, 14, 10, 13, 7, 15, 18, 8, 17, 6)$$

4.5.6 Augmented designs

A final aspect of block designs for comparing genotypes is the addition of unreplicated genotypes to a replicated genotype trial. The additional genotypes are likely to be promising alternatives deserving an initial assessment alongside the main genotypes under current consideration. Specialized designs are not needed since comparisons of added genotypes with the main genotypes in the trial cannot be very precise because of the lack of replication. The major danger in adding genotypes is the potential for inflating σ^2 by increasing block size, and for this reason we suggest that no more than two extra genotypes be added to each block (see also Section 6.2.3).

4.6 PRECISION

Section 4.4.2 considered the precision of comparisons between genotypes, based on indirect comparisons from incomplete blocks. In general, the precision of estimates of yield differences between genotypes for any experimental design is described by the standard error of a difference between two genotype means (SED):

$$\text{SED} = \sqrt{\left(2C\sigma^2\right)}$$

The SED has two components. The basic variance between plots within blocks, σ^2, is a characteristic of the particular plots and of the division of the plots into blocked groups. With sensible blocking we expect σ^2 to decrease with the number of plots per block.

The second component, the factor C, represents the effectiveness of comparisons in the design. In complete block designs, or in designs without blocking, C will be $1/r$ where r is the number of replicates per genotype. When using incomplete blocks some information is lost because some direct comparisons are replaced by indirect comparisons. Consequently C will be larger than $1/r$. The amount by which C is greater than $1/r$ depends on the particular design pattern, and also on the particular genotype comparisons. Often the variation of C values for different comparisons within a design is small and we can use an average C value for all genotype comparisons.

The design efficiency for a design, or for a particular comparison of genotypes, is defined as

$$E = 1/(rC)$$

The average design efficiency for an incomplete block design is always less than one because indirect comparisons are used instead of direct comparisons. However, the average values of E are much higher than might be expected from considering only direct comparisons.

Using incomplete block designs in place of complete block designs will normally reduce σ^2 substantially below the value for complete blocks, σ_0^2. In an extensive investigation of UK cereal trials, Patterson and Hunter (1983) found that σ^2 increased exponentially with the block size, and that using incomplete rather than complete blocks reduced the average σ^2 by over 40%.

The important practical question is whether the standard errors of comparisons of genotypes are smaller for trials designed in incomplete rather than complete blocks. The answer depends on the balance between the gain from reducing σ^2 and the loss from design inefficiency. We define the practical measure of experiment efficiency as the ratio of the average variance of the genotype differences for complete versus incomplete blocks:

$$E^* = \frac{\sigma_0^2}{r} / C\sigma^2 = \left[\frac{1}{rC}\right]\left[\frac{\sigma_0^2}{\sigma^2}\right]$$

If this is greater than 1.0, an incomplete block design will give an overall improvement of precision. Investigation of previous trial results indicates that using incomplete block designs does give improvements of experiment efficiency. For UK cereal trials, Patterson and Hunter (1983) quote average efficiency gains of 30%.

4.6.1 Approximate design efficiencies for incomplete block designs

We can predict an approximate design precision for an incomplete block design by considering the average number of direct comparisons for each genotype pair, λ, and the average replication per genotype, r. The smallest possible average variance for genotype differences, based on all observations per genotype, is

$$\text{MINVAR} = 2\sigma^2/r$$

The largest possible average variance for genotype difference, based on using direct comparisons, is

$$\text{MAXVAR} = 2\sigma^2/\lambda$$

A first approximation to the average variance of genotype differences is then

$$\text{VAR} = \text{MINVAR} + \frac{\text{MAXVAR} - \text{MINVAR}}{g}$$

where g is the number of genotypes.

This variance formula is exact for the class of balanced incomplete block designs but slightly underestimates the variances and standard errors for the unbalanced designs used with larger numbers of genotypes. To illustrate the degree of approximation, Table 4.13 gives the actual standard errors for each of the seven designs considered earlier in this chapter, together with MINVAR, MAXVAR, VAR and √VAR. The standard error calculated from the approximate formula is within 4% of the mean SED in all cases. The design efficiency, E, is MINVAR/(SED)^2.

Another important feature of Table 4.13 is the relatively narrow range of standard errors for each design. This may seem surprising for the design

Table 4.13. Standard error (SED) and approximate variance of genotype differences (for $\sigma^2 = 1$) for seven incomplete block designs with g genotypes in s blocks of k plots

Table reference and design parameters (g s k)	*SED*		*√VAR*	*VAR*	*MINVAR*	*MAXVAR*
Table 4.6	min	0.732				
(6 6 4)	mean	0.748	0.743	0.556	1/2	5/6
	max	0.760				
Table 4.7	all	0.756	0.756	0.571	1/2	1
(7 7 4)						
Table 4.8	min	0.816				
(12 6 6)	mean	0.872	0.856	0.733	2/3	22/15
	max	0.919				
Table 4.9	min	0.894				
(25 15 5)	mean	0.913	0.894	0.800	2/3	4
	max	0.931				
Table 4.10	min	0.924				
(28 21 4)	mean	0.963	0.926	0.857	2/3	6
	max	0.986				
Table 4.11	min	0.897				
(50 30 5)	mean	0.938	0.904	0.817	2/3	49/6
	max	0.961				
Table 4.12	min	0.800				
(18 20 3/4)	mean	0.822	0.806	0.650	1/2	51/16
	max	0.856				

with 50 genotypes (Table 4.11), where the proportion of pairs of genotypes which have any direct comparison is only 25%. Again we see the power of indirect comparisons. In Section 4.5.2 we discussed precision for the third design of Table 4.13 and noted that the actual design used (in Table 4.8) could have been slightly improved. We can now reveal that the mean and maximum SEDs change only by 0.1% for the improved design. Indeed, if we consider the design of Table 4.14, which is apparently much worse than Table 4.8 with very similar divisions of genotypes into blocks for the three replicates, the mean SED only increases from 0.856 to 0.900. The moral is that designs have to be very bad before any substantial loss of precision results.

4.7 MULTIPLE LEVELS OF INFORMATION

4.7.1 Information from variation between block totals

So far we have considered only the information derived from differences between plot yields in each block. However, the differences between incomplete block totals also contain information about the differences between genotypes.

We will consider a simple example with six genotypes *A* to *F*, in ten blocks of three plots. Suppose that the yields from the individual plots are lost and only the total yield for each block is available. The block yields with genotypes contributing to each total are given in Table 4.15. Some patterns can be seen fairly quickly: genotype *A* occurs in five of the six highest yielding blocks; genotype *C* occurs in the lowest yielding blocks, except when it occurs with genotype *A*: patterns for genotypes *B*, *D*, *E* and *F* are less clear but genotype *B* tends to appear in higher yielding blocks and genotype *F* in lower yielding blocks. Using the least squares method to estimate the genotype yields which give the best agreement with the block totals, we obtain: *A* 21.5; *B* 15.5; *C* 2.5; *D* 12.9; *E* 9.3; *F* 7.5. The estimated genotype differences confirm our initial patterns with genotype *A* clearly giving the largest estimated yield, genotype *C* clearly below the rest and genotypes *B*, *D*, *E* and *F* showing smaller differences. The extent to which the variation between the block

Table 4.14. A bad design for comparing 12 genotypes in blocks of six plots using three replicates

Replicate 1		*Replicate 2*		*Replicate 3*	
Block I	*Block II*	*Block III*	*Block IV*	*Block V*	*Block VI*
1	3	2	1	2	1
2	4	4	3	3	4
6	5	6	5	6	5
7	8	7	8	7	8
11	9	10	9	11	9
12	10	12	11	12	10

Table 4.15. Between-block analysis for incomplete block trial with six genotypes in 10 blocks showing genotype design, block totals and analysis of variance of block means

Blocks	*Genotypes included*	*Observed totals*	*Fitted totals*
I	*A, B, C*	39	39.5
II	*A, B, D*	48	49.9
III	*A, C, E*	34	33.3
IV	*A, D, F*	45	41.9
V	*A, E, F*	37	38.3
VI	*B, C, F*	28	25.5
VII	*B, D, E*	39	37.7
VIII	*B, E, F*	31	32.3
IX	*C, D, E*	25	24.7
X	*C, D, F*	20	22.9

Source	SS	df	MS	*F*
Genotypes	660	5	132.0	15.5
Residual	34	4	8.5	
Total	694	9		

SS, sum of squares; df, degrees of freedom; MS, mean square, *F*, *F*-ratio.

totals can be explained by the differences between genotypes may be displayed in an analysis of variance for block totals (Table 4.15). This shows that the sum of squares of the differences between the observed and fitted block totals is 33.8, compared with the total sum of squares of 694 for the original ten totals.

4.7.2 Using information from multiple levels

The preceding example shows that differences between block totals can provide information about genotype differences. This information can supplement that obtained from the more usual analysis of variance between plots within blocks. The two sources of information about genotype differences are quite separate and we now consider how best to combine the two sets of estimates.

First we consider how the two sets of estimates arise and derive their relative precisions. When discussing incomplete block designs (Section 4.6), we used the concept of design efficiency to quantify the information lost through using indirect rather than direct comparisons. The design efficiency, $E = (1/rC)$, can also be viewed as describing the proportion of the information about genotype differences which is retained in the comparisons between plot yields within blocks. The information, $1 - E$, which is 'lost' is precisely that provided from the comparisons between block totals. For the example in Section 4.7.1, each of the $g = 6$ genotypes occurs $r = 5$ times, each genotype pair occurs $\lambda = 2$ times and the variance factor (Section 4.6.1) for the balanced design is

$$\frac{\text{VAR}}{\sigma^2} = \frac{2}{5} + \frac{\frac{2}{2} - \frac{2}{5}}{6} = \frac{1}{2}$$

The design efficiency $E = 0.4/0.5$ (= MINVAR/VAR) or 80%. The remaining 20% of the information is contained in the differences between the block totals.

However, the actual information in the two levels is determined, not only by the design efficiency, but also by the random variation in the two levels. If the blocking has been well arranged, the variation of plots within blocks σ^2 will be small, and correspondingly the variation between blocks will be large. The extent to which plots in different blocks vary more than plots in the same block is measured by the additional block component of variance, σ_b^2. The relative amounts of information in the within-block and between-block levels can be shown to be

$$\frac{E}{\sigma^2} : \frac{1 - E}{\sigma^2 + k\sigma_b^2}$$

where k is the block size. A high proportion (usually over 90%) of the information about the genotype differences is generally contained in the within-block level, and sometimes the additional extraction and combination of information from the between-block level are ignored.

In many trials, however, it will be worthwhile using both sets of estimates. How can these two estimates be combined? A general statistical result tells us that if we have two estimates (d_1, d_2) of the same quantity, then the most precise single estimate is a weighted average of the two estimates, the weights being the reciprocals of the variances (s_1^2, s_2^2) of the two estimates:

$$\left[\frac{d_1}{s_1^2} + \frac{d_2}{s_2^2}\right] \Big/ \left[\frac{1}{s_1^2} + \frac{1}{s_2^2}\right]$$

4.7.3 Multiple level estimation by REML

The estimation of variety effects from both the within-block and between-block levels of information has been presented as two separate operations with a subsequent combination of information. This emphasized the distinction between the two sources of information. However, since the data for the two levels derives from the same set of measurements, the analyses can be combined in a single procedure. This is possible through the method known as 'residual maximum likelihood' or REML, introduced by Patterson and Thompson (1971) and available within the GENSTAT, SAS and BMPD packages (Brown and Kempton, 1994). An extensive description of the REML method and its application is given by Patterson *et al.* (1990).

REML models the different levels of variation (within and between blocks) using variance components. The estimation of the two variance components is complicated by the differences between genotypes which are regarded as fixed effects and are the principal objectives for the estimation procedure. The estimation procedure is iterative with two alternating stages: (i) using estimates of genotype differences, estimate the variance components; (ii) using estimates of the two variance components, estimate the genotype differences. Initial estimates of either genotype differences or variance components are required to start the procedure – the usual practice is to estimate initial genotype differences from the within-block variation, as described in Section 4.4.

REML can be generalized to analyse experimental structures with more than two levels of variation. If the experimental variation can be split into a number of levels or strata then, by estimating the variance components in these strata, we can obtain information about genotype differences using the combination of the separate information from the separate strata. The procedure is illustrated by a two-replicate trial for 35 wheat genotypes (Patterson *et al.*, 1990). Each replicate is arranged in five rows by seven columns, and the design uses both row and column blocking systems (see Section 4.8). The REML analysis combines information from the differences between rows, the differences between columns and the differences within rows and columns to estimate genotype differences.

The design and plot yields are given in Table 4.16 – yields for two plots were missing. In defining the model for analysis by REML we must decide which effects will be considered as random and which as fixed. Thus, if rows are taken as fixed, REML will estimate the individual differences between rows, whereas with random rows REML estimates only the variance of the population of effects from which the sample of rows is drawn. We could assume fixed or random effects for the sets of replicates, rows, columns and genotypes. In this case, genotypes are chosen as fixed effects because we are interested in making a comparison between the particular 35 selections. Replicates are also taken as fixed effects since there are only two replicates and thus only one degree of freedom (df) for estimating the variance of replicate effects; this will not give a useful variance estimate.

Table 4.16. Design and yield data for 35 genotypes of wheat in two replicates with five rows and seven columns

Design							*Yields*						
Replicate 1													
20	*4*	*33*	*28*	*7*	*12*	*30*	3.77	3.21	4.55	4.09	5.05	4.19	3.27
10	*14*	*16*	*21*	*31*	*6*	*18*	3.44	4.30	–	3.86	3.26	4.30	3.72
22	*11*	*19*	*26*	*29*	*15*	*23*	3.49	4.20	4.77	2.56	2.87	1.93	2.26
24	*25*	*5*	*32*	*2*	*27*	*8*	3.62	4.52	4.23	3.76	3.61	3.62	4.01
17	*9*	*3*	*34*	*13*	*35*	*1*	3.81	3.75	4.81	3.69	4.61	2.68	4.15
Replicate 2													
31	*19*	*25*	*34*	*20*	*8*	*6*	4.70	7.37	5.03	5.33	5.73	4.70	5.63
24	*21*	*12*	*4*	*23*	*13*	*3*	4.07	5.66	4.98	4.04	4.27	4.10	4.75
11	*7*	*26*	*5*	*35*	*10*	*30*	5.66	6.43	4.59	5.20	4.83	4.70	4.23
33	*9*	*17*	*18*	*32*	*15*	*2*	5.71	6.13	4.63	5.48	5.47	–	4.16
1	*27*	*16*	*29*	*14*	*28*	*22*	5.22	6.16	4.20	4.66	5.54	3.81	3.60

Assuming row and column effects are fixed, we can estimate the genotype differences by least squares analysis, adjusting for row and column effects. For this analysis the minimum, mean and maximum standard errors of genotype differences are based on the estimate of σ^2, $s^2 = 0.088$:

minimum 0.360 mean 0.408 maximum 0.581

In contrast, if we assume that row and column effects are random then we estimate three variance components:

for plots $s^2 = 0.090\ (\pm\ 0.037)$
for rows $s_r^2 = 0.064\ (\pm\ 0.048)$
for columns $s_c^2 = 0.192\ (\pm\ 0.097)$

and obtain standard errors for genotype differences:

minimum 0.351 mean 0.385 maximum 0.536

The random effects model gives a reduction of about 5% in the average standard error (about 11% reduction in variance) indicating that about 11% of the information about genotype differences is provided by differences between rows and between columns. The biggest change is in the maximum SED, so the range of SEDs is also reduced. The estimates of precision for row and column variance components (based on 8 df and 12 df respectively) suggest that both are larger than zero.

If we decided that the row variance component was small enough to ignore we could fit a model with fixed genotype and replicate effects and with random column effects (and no row effects). The estimates from REML are

for plots $s^2 = 0.167\ (\pm\ 0.053)$
for columns $s_c^2 = 0.164\ (\pm\ 0.100)$

and the standard errors for genotype differences are

minimum 0.437 mean 0.476 maximum 0.661

Hence, failing to include the variation between rows in any form gives much less precise estimates. For this example, the most appropriate model is that with random row and column effects and fixed genotype and replicate effects.

REML provides a fully efficient analysis, but two aspects of fitting models are not explicit in the results. First, we should check that the degrees of freedom on which variance components are estimated are sufficient for reliable estimates; a minimum of 8 df is a fairly safe limit. Second, the process of selecting a model itself introduces a small additional imprecision into the estimation procedure, which is not reflected in the SEDs from the REML analysis.

4.8 ROW AND COLUMN DESIGNS

It is indisputable that the information from large genotype trials will be greatly improved by using experimental designs with small, incomplete blocks. The construction of incomplete block designs is now relatively simple and the alpha methodology provides a well-understood system for producing appropriate and efficient designs. A computer program, ALPHA+, is also available to provide designs with the highest possible efficiency, within the class of resolvable designs.

Analysis of data from past genotype trials suggests that allowing for block differences in two directions can further improve precision. The example of a REML analysis in Section 4.7 illustrates the gains possible from adjusting plot yields for row blocks and column blocks. Kempton, Seraphin and Sword (1994) provide more substantial evidence.

To benefit most from the row and column analysis, it is important to design efficient experiments with row and column blocking systems. Among the different forms of row and column designs, the simplest is where each replicate consists of a rectangular array of M rows by N columns. In practice, such designs are suitable when the number of genotypes is not too large, so that both M and N are no larger than about eight plots. One classical form of design, the lattice square, is available when the number of genotypes is the square of an integer, k, so that $M = N = k$. If k is an odd number, then a lattice square design with $(k + 1)/2$ replicates exists and is balanced (in the sense of equal precision of genotype comparisons when row and column variance components are equal). An example for 25 genotypes in three replicates, each consisting of five rows by five columns, is shown in Table 4.17. For each genotype, all the other genotypes occur either once in the same row or once in the same column. A second set of replicates could be added with rows and columns interchanged. However, using more than three or four replicates will rarely be appropriate for a single design. If k is even, balanced lattice square designs exist for $k = 4$ and $k = 8$ with five and nine replicates respectively.

More generally, good row and column designs should have (i) an efficient design in the row-blocks; (ii) an efficient design in the column-blocks; and (iii) the two separate designs complementary. Thus in the lattice square of Table 4.17, the row-block design has each pair of genotypes together in a block either once or not at all. The same is true for the column-block design so each design is efficient in itself. The two designs are complementary in that each genotype pair not occurring in a row-block does occur in a column-block.

A natural approach to constructing row–column designs with a single rectangle of plots for each replicate would be to find separate alpha designs for rows and columns and then combine designs which are complementary in the sense of avoiding genotype pair repeats in rows and columns. As noted in Section 4.5.4, alpha designs are particularly effective when the block size, k, is less than $\sqrt{g}$, the square root of the number of genotypes. However, for rows and columns in a rectangle, it is impossible for both row size and column size to be less than $\sqrt{g}$. We therefore start with an alpha design for the smaller block size (which is the more important blocking system for efficiency) and attempt to rearrange the genotypes into an effective second blocking system. As an example, we rearrange the alpha design of Table 4.10 in three replicate rectangles of four columns by seven rows in Table 4.18. Inevitably some genotype pairs are repeated in the longer columns, there are some genotype pairs which are together in both a row and a column, and some genotypes pairs never occur together: for example, genotype *1* occurs with genotypes *5*, *7* and *16* in two columns; with genotypes *11*, *22* and *27* in both rows and columns; and never with genotypes *10*, *13*, *19*, *23*, *24* and *28*. The design can be improved, but is definitely better than simply randomizing the genotypes within rows before fitting row and column effects. To achieve a better design, computer search techniques are required. An algorithm by Nguyen and Williams (1993) is incorporated in the latest version of ALPHA+.

For larger numbers of genotypes, each replicate may be split into two or more row by column rectangles, giving smaller block sizes in both rows and

Table 4.17. A lattice square design for 25 genotypes in three replicates

Replicate 1						*Replicate 2*						*Replicate 3*					
Rows	*Columns*					*Rows*	*Columns*					*Rows*	*Columns*				
	1	2	3	4	5		6	7	8	9	10		11	12	13	14	15
I	*1*	*2*	*3*	*4*	*5*	VI	*1*	*10*	*14*	*18*	*22*	XI	*1*	*8*	*15*	*17*	*24*
II	*6*	*7*	*8*	*9*	*10*	VII	*23*	*2*	*6*	*15*	*19*	XII	*25*	*2*	*9*	*11*	*18*
III	*11*	*12*	*13*	*14*	*15*	VIII	*20*	*24*	*3*	*7*	*11*	XIII	*19*	*21*	*3*	*10*	*12*
IV	*16*	*17*	*18*	*19*	*20*	IX	*12*	*16*	*25*	*4*	*8*	XIV	*13*	*20*	*22*	*4*	*6*
V	*21*	*22*	*23*	*24*	*25*	X	*9*	*13*	*17*	*21*	*5*	XV	*7*	*14*	*16*	*23*	*5*

Table 4.18. An alpha design to compare 28 genotypes rearranged in four columns by seven rows for each replicate

Replicate 1					*Replicate 2*					*Replicate 3*				
Rows	*Columns*				*Rows*	*Columns*				*Rows*	*Columns*			
	1	2	3	4		5	6	7	8		9	10	11	12
I	*1*	*8*	*15*	*22*	VIII	*1*	*9*	*17*	*26*	XV	*1*	*11*	*21*	*27*
II	*2*	*9*	*16*	*23*	IX	*27*	*2*	*10*	*18*	XVI	*12*	*2*	*23*	*17*
III	*3*	*10*	*17*	*24*	X	*11*	*19*	*28*	*3*	XVII	*22*	*15*	*3*	*13*
IV	*4*	*11*	*18*	*25*	XI	*20*	*22*	*4*	*12*	XVIII	*16*	*28*	*14*	*4*
V	*5*	*12*	*19*	*26*	XII	*5*	*21*	*23*	*13*	XIX	*18*	*8*	*5*	*24*
VI	*6*	*13*	*20*	*22*	XIII	*14*	*6*	*15*	*24*	XX	*25*	*6*	*19*	*9*
VII	*7*	*14*	*21*	*28*	XIV	*16*	*25*	*8*	*7*	XXI	*7*	*26*	*10*	*20*

columns. Where the replicates are adjacent in the field, row or column blocks can run across the replicates. For the example of Table 4.11 it was found that 'row' blocks of six plots, with each block including two plots from each replicate, improved the efficiency. In the original design of Table 4.11 several genotypes are duplicated in the row blocks and genotype pairs in column blocks are repeated in the row blocks. Interchanging the column blocks within replicates and the genotypes within column blocks can produce considerable improvements. Designs with this additional property are described as latinized (Williams, 1986b).

4.9 TRIAL DESIGNS FOR SPATIAL ANALYSIS

Chapter 5 describes recently developed spatial methods of analysis for field plot data, which are based on the observed similarity of yields from adjacent plots. If a spatial method of analysis is to be used, how should this influence the choice of experimental design?

Methods of spatial analysis get most information on genotypes from the differences in yield between adjacent plots. Therefore the most important design characteristic is that no pair of genotypes should occur on adjacent plots more than once. The designs may be constructed specifically to avoid repeats of adjacent pairs, and investigations have been carried out of designs with neighbour balance (Gleeson and Eccleston, 1992). However, since the pairing of plots has similarities with using small blocks, incomplete block designs, which are readily available, should produce sensible solutions to the design problem for spatial analysis. It therefore seems reasonable (B.R. Cullis, personal communication) to use incomplete block designs with small block sizes (say four or five plots per block) where spatial analysis may be applied, used either as a supplement or as an alternative to block analysis.

4.10 MULTILOCATION EXPERIMENTS

When a set of experiments are planned for multiple locations, we have to consider both the individual design for each location and the overall set of designs for the combined locations. In many cases, the set of genotypes will be identical at all locations, but sometimes a core set of genotypes is tested at all locations with different additional genotypes added at different locations.

Ideally, we would like to choose an appropriate block size, or sizes, and appropriate sets of plots for each block at each location separately. When available, local knowledge about patterns of plot-to-plot variation should provide the potential for improving precision through detailed planning of the layout and blocking pattern. However, using smaller block sizes and alpha designs reduces the benefits from tailoring each design to local conditions. We therefore recommend selecting a suitable small-block alpha design and using it, with different randomizations, for all locations. The local trials officers will then use their knowledge of local conditions to choose the best trial layout at each location. It is important to ensure that different randomizations are used for each location so that any combined analysis is valid.

If each location has a common core of genotypes and additional local genotypes, then an alpha design should be selected for the core genotypes and blocks augmented with the additional local genotypes (Section 4.5.6). In some cases it may be appropriate to incorporate the local genotypes as an additional set when constructing the alpha design (Section 4.5.4).

Another area of decision in planning a multilocation experiment is how to choose the locations. This choice should usually have both systematic and random components. The systematic requirement will ensure that the combined experiment adequately represents the intended range of environments. This range of environments, which is sometimes called the recommendation domain for the chosen genotypes, should be defined at the start of the selection programme and trial locations should include the extreme conditions within this domain. Ideally, some locations should also be selected at random from sub-populations of available locations, each sub-population representing a particular part of the recommendation domain. Choice of locations is discussed further in Chapter 9.

4.11 ADDITIONAL FACTORS

In many genotype trials, it may be valuable to investigate the extent to which genotypes respond differently to changes in the management practice of the trial. This is a major motivation for using multilocation experiments but different locations represent different environments in a rather imprecise way. Including an additional treatment factor in a single-location experiment allows an experimentor to examine the extent of interaction between the genotype differences and the levels of the treatment factor, and to obtain more broadly based estimates of the differences between the genotypes.

The ideal way to introduce additional factors into a single-location genotype trial is to apply different levels of the treatment factor to whole replicates. It is usually desirable to retain two identically treated replicates to provide an unbiased estimate of σ^2. The main effect differences between the levels of treatment applied to the whole replicates are not estimable and are not important in this context. The interaction sum of squares for genotypes $\times$ applied treatment factor is removed from the error sum of squares, but provided there are two identical replicates and at least fifteen genotypes in the trial there will be no shortage of degrees of freedom for the error sum of squares.

5
Spatial analysis

A.C. GLEESON

5.1 WHY SPATIAL ANALYSIS?

A central problem confronting a plant breeder when comparing genotypes in a field trial is that the yield of a genotype is markedly affected by the condition, particularly the soil moisture and fertility, of the plot in which the genotype is sown. Comparison problems increase with separation between plots. For example, Batchelor and Reed (1918), in a study on the variation of orange tree yields, noted that 'marked soil variations occur which tend to make adjacent tree or plot yields alike'. Fisher (1935), in advocating the importance of replication and randomization in agricultural field trials, observed that 'after choosing an area, we usually have no guidance beyond the widely verifiable fact that patches in close proximity are commonly more alike, as judged by the yield of crops, than those which are further apart'.

The non-uniformity of conditions over the trial is sometimes referred to as within-trial heterogeneity or 'fertility' trends. The most common approach to overcome the problems of within-trial heterogeneity is to group plots into small blocks for which the assumption of uniform conditions within blocks is more realistic. Blocking, combined with replication and randomization, can greatly increase the precision of treatment estimates (see Chapter 4). Following a pioneering discussion paper by Wilkinson *et al.* (1983), an alternative approach of applying 'spatial' or 'neighbour' methods to the analysis of agricultural field trials attracted much interest.

A concern over using spatial analysis for field trials is that 'treatment estimates and their errors rely on the chosen (error) model which may not reflect the real situation' (Besag and Kempton, 1986). It is therefore important to check the adequacy of any spatial model for the data being analysed.

5.2 SPATIAL AUTOCORRELATION

Heterogeneity or fertility trends result in correlation between plots, so a useful statistic for examining the within-trial heterogeneity is the spatial

Statistical Methods for Plant Variety Evaluation. Edited by R.A. Kempton and P.N. Fox. Published in 1997 by Chapman & Hall, London. ISBN 0 412 54750 3.

autocorrelation coefficient. If we consider a rectangular array of field plots comprising M ranges of N plots (see Fig. 2.1) and denote the sample correlation coefficient between plots u ranges and v plots apart by $r(u,v)$, then $r(0,1)$ is the correlation between adjacent plots in a range [Fig. 5.1(a)] and $r(0,2)$ is the correlation between plots two apart in a range [Fig. 5.1(b)]. Similarly $r(0,1)$ and $r(0,2)$ are the correlations between plots in the same position, or pass, in adjacent ranges and two ranges apart [Fig. 5.1(c) and (d)].

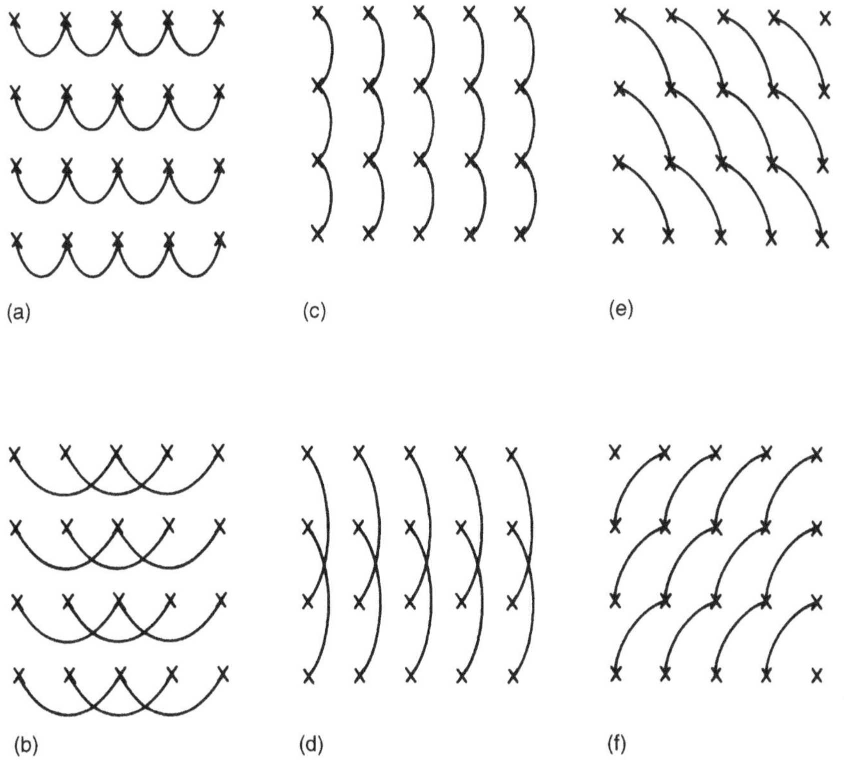

Fig. 5.1. Plot pairs used to derive spatial correlations at different spatial separations. (a) $r(0, 1)$; (b) $r(0, 2)$; (c) $r(1, 0)$; (d) $r(2, 0)$; (e) $r(1, 1)$; (f) $r(1, -1)$.

Now consider the sample correlations in different ranges and passes. For example, $r(1,1)$ can be considered as the sample correlation between plots and their south-east neighbours [Fig. 5.1(e)], whereas $r(1,-1)$ is the sample correlation between plots and their north-east neighbours [Fig. 5.1(f)]. In spatial analysis it is often sufficient to consider the average sample correlation of plots u ranges and v plots apart which we define as

$$\bar{r}(u,v) = \tfrac{1}{2}\,[r(u,v) + r(u,-v)]$$

When there are different genotypes in each plot, spatial correlations of yield data provide limited information unless the yields are first treatment-corrected.

A crude method of doing this in a replicated trial is to subtract the relevant genotype mean from each plot yield.

To define the sample spatial autocorrelation $\bar{r}(u,v)$, we denote the treatment-corrected yield of the jth plot ($j = 1, \ldots, N$) in the rth range ($r = 1, \ldots, M$) by y_{rj}. Thus the spatial arrangement of plot yields in the field is

$$\begin{array}{ccccc} y_{11} & y_{12} & y_{13} & \cdots & y_{1N} \\ y_{21} & y_{22} & y_{23} & \cdots & y_{2N} \\ y_{31} & y_{32} & y_{33} & \cdots & y_{3N} \\ \cdot & \cdot & \cdot & & \cdot \\ \cdot & \cdot & \cdot & & \cdot \\ \cdot & \cdot & \cdot & & \cdot \\ y_{M1} & y_{M2} & y_{M3} & \cdots & y_{MN} \end{array}$$

We define the matrix of sample covariances between plots u ranges apart and v plots apart as

$$C_{u,v} = [(M - |u|)(N - |v|)]^{-1} \Sigma\Sigma\, y_{rj}\, y_{r+u,\, j+v}$$

and

$$C_{u,-v} = [(M - |u|)(N - |v|)]^{-1} \Sigma\Sigma\, y_{r,j+v}\, y_{r+u,j}$$

where the double summation is over $r = 1, \ldots, M - |u|$ and $j = 1, \ldots, N - |v|$, and $|u|$ denotes the absolute value of u. The sample covariance $C_{0,0}$ is the variance of the y_{rj}, so the sample autocorrelations is $r(u,v) = C_{u,v}/C_{0,0}$ and $r(u,-v) = C_{u,-v}/C_{0,0}$.

The matrix of spatial correlations for plot separations (or 'lags') u and v (Table 5.1) provides the experimenter with a quick appreciation of the pattern of heterogeneity within a trial. As the lags (either u or v) increase, the number of pairs over which the correlation is calculated decreases, hence the accuracy of the sample correlation will also decrease. In practice, therefore, we usually consider only sample correlations for small lags, i.e. the top left corner of the correlation matrix (Table 5.1).

Table 5.1. Matrix of spatial correlations

Distance apart (ranges)	*Distance apart (plots within a range)*				
	0	1	2	3	...
0	1.00	$r(0,1)$	$r(0,2)$	$r(0,3)$	...
1	$r(1,0)$	$\bar{r}(1,1)$	$\bar{r}(1,2)$	$\bar{r}(1,3)$	...
2	$r(2,0)$	$\bar{r}(2,1)$	$\bar{r}(2,2)$	$\bar{r}(2,3)$	...
3	$r(3,0)$	$\bar{r}(3,1)$	$\bar{r}(3,2)$	$\bar{r}(3,3)$	...
...	...	...	...	...	...

5.3 SAMPLE SPATIAL CORRELATIONS FOR UNIFORMITY TRIALS

The level of heterogeneity, trend or correlation at a location is most clearly seen from a uniformity trial, where only one genotype is grown but plot size spacing and husbandry are otherwise the same as for a genotype selection trial. To reduce costs, a uniformity trial may be superimposed upon the seed multiplication lot of an advanced selection.

5.3.1 Example 1: orange trees

Batchelor and Reed (1918) examined the yields from a uniformity trial on 1000 orange trees planted in a 20 × 50 array with 7 m spacing in each direction. The spatial correlations of the combined two-year yields for trees up to five rows and columns apart are presented in Table 5.2.

Table 5.2. Spatial correlations among yields of orange trees (Example 1)

Distance apart (rows)	*Distance apart (trees within rows)*					
	0	*1*	*2*	*3*	*4*	*5*
0	1.00	0.51	0.41	0.38	0.30	0.28
1	0.52	0.41	0.34	0.32	0.28	0.25
2	0.47	0.37	0.31	0.29	0.27	0.24
3	0.43	0.31	0.26	0.27	0.23	0.21
4	0.39	0.30	0.25	0.24	0.21	0.18
5	0.35	0.28	0.23	0.22	0.20	0.17

If a location is uniform, all of the spatial correlations should be zero (except $r(0,0)$ which is always 1). To test this, we compare the sample correlations with twice their standard error. A crude estimate of the standard error is the reciprocal of the square root of the sample size, in this case, $1/\sqrt{1000} \approx 0.03$. Hence there is significant correlation in the yields of neighbouring trees. The correlation decreases as the distance between the trees increases, and the rate of decrease is about the same along the rows as between the rows. This is to be expected with a square planting pattern, unless there is a systematic trend across the trial.

5.3.2 Example 2: wheat plots of 9 m × 1.8 m

This wheat uniformity trial consisted of 180 plots of size 9 m × 1.8 m in six ranges of 30 plots. The trial covered an area of 54 m × 54 m. The sample spatial correlations for plot yield are given in Table 5.3.

The approximate standard error of the spatial correlations is $1/\sqrt{180} \approx 0.07$, so the only substantial correlations are between the plots within a range, and these decrease with the separation between the plots. This pattern is not unexpected since the distance between adjacent plot centres between ranges is five times that within a range. One dimensional spatial analysis is clearly adequate.

Table 5.3. Spatial correlations among yields of wheat plots (Example 2)

Distance apart (ranges)	*Distance apart (plots within ranges)*					
	0	*1*	*2*	*3*	*4*	*5*
0	1.00	0.62	0.42	0.34	0.25	0.22
1	0.09	0.06	0.05	0.07	0.09	0.08
2	0.01	0.01	0.04	0.04	0.04	0.00

5.3.3 Example 3: wheat plots of 20 m × 1.8 m

This wheat uniformity trial used larger (20 m × 1.8 m) plots in eight ranges of 20 plots and covered 160 m × 36 m. Table 5.4 gives the spatial correlations for plot yields.

Table 5.4. Spatial correlations among yields of wheat plots (Example 3)

Distance apart (ranges)	*Distance apart (plots within ranges)*					
	0	*1*	*2*	*3*	*4*	*5*
0	1.00	0.74	0.61	0.49	0.47	0.37
1	0.28	0.22	0.22	0.17	0.19	0.18
2	0.15	0.13	0.11	0.09	0.09	0.09
3	0.14	0.10	0.08	0.05	0.05	0.02

The approximate standard error of the spatial correlations is $1/\sqrt{160} \approx 0.08$. The correlations between the plots are higher than for Example 2 in both directions and the correlation between the ranges is statistically significant. Despite the rectangularity of the plots (plot length to width ratio is greater than ten) a two-dimensional spatial analysis may be required.

5.4 MODELS FOR THE ANALYSIS OF FIELD TRIAL DATA

A model for the analysis of field trial data was introduced in Section 4.2.2 as

$$y = \mu + b + g + e \tag{5.1}$$

where it is assumed that plots are grouped into homogeneous blocks and that genotype effects g and block effects b are additive. The plot error e is defined as the residual after fitting all of the other terms in the model. We will illustrate the different assumptions of various analyses by describing each component of the model.

In this section we consider uniformity data where genotype effects are omitted from the model. The yield of the jth plot in the rth range is then

$$y_{rj} = \mu + b_r + e_{rj} \tag{5.2}$$

where the blocks are initially identified with ranges and individual plot errors e_{rj} (i.e. plot yields adjusted for range means) are independently and identically distributed Normal deviates with mean zero and constant variance [i.i.d. $N(0,\sigma^2)$].

Now consider the fit of Equation 5.2 to the uniformity data in Example 2. If the plot errors are i.i.d. $N(0,\sigma^2)$, their sample lag 1 correlation (i.e. the correlation of errors from adjacent plots) should not deviate significantly from zero. The plot values for range 1 are illustrated in Fig. 5.2. From Table 5.3, the lag 1 autocorrelation, over all ranges, is 0.62 with an approximate standard error of 0.07, so the assumption of independence does not hold.

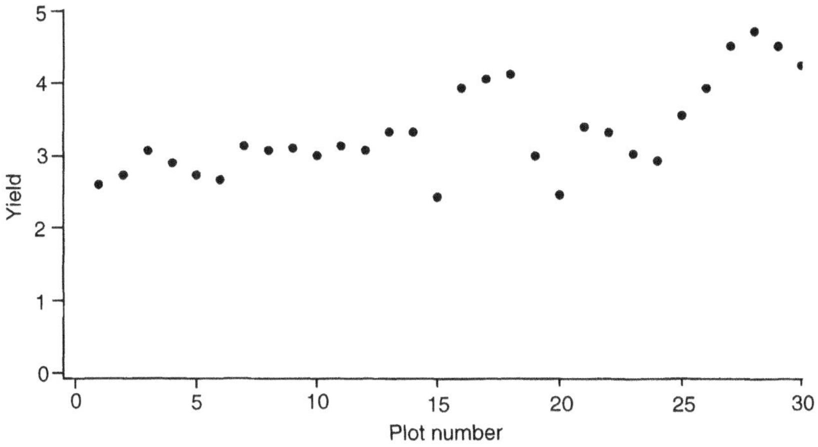

Fig. 5.2. Plot yields for range 1 of Example 2.

Suppose that instead of imposing a complete block design on this trial, we use an incomplete block design, for example an alpha design with six blocks of five plots per range (Section 4.5.4). The model, omitting genotype effects, is

$$y_{rj} = \mu + b_{rj} + e_{rj} \tag{5.3}$$

where b_{rj} is now the block effect associated with the jth plot in the rth range. The effects b are either fixed, for an intra-block analysis, or random, for an incomplete block analysis with the recovery of inter-block information (see Section 4.7); the errors e are assumed i.i.d. $N(0,\sigma^2)$.

Figure 5.3 depicts the estimated block effects b and the errors e, in field plot order for range 1 in Example 2. The lag 1 autocorrelation of the errors is 0.03, with a standard error of 0.07, which shows the effectiveness of blocking in reducing the level of spatial correlation of the errors.

Spatial analysis involves describing the plot variation within ranges for Equation 5.2 by 'trend plus error' terms. Omitting genotype effects, the equation can be written as

$$y_{rj} = \mu + \xi_{rj} + e_{rj} \tag{5.4}$$

where ξ is the trend and e the new residual error.

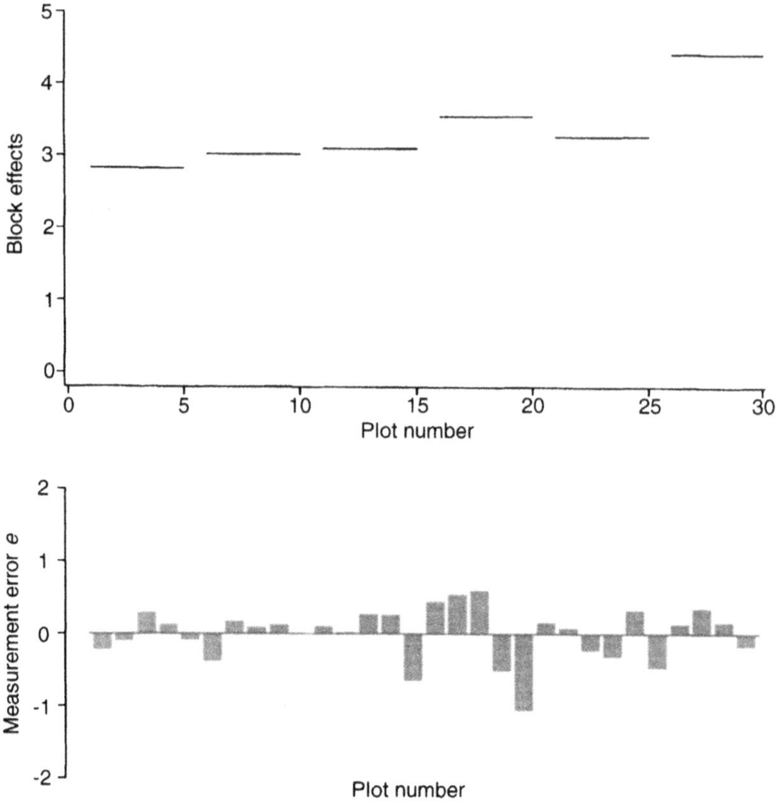

Fig. 5.3. Block effects and plot errors for Example 2.

Zimmerman and Harville (1991) modelled ξ by a low-order polynomial for the plot position. If in Example 2 we assume a quadratic polynomial for ξ, the trend and errors for the range 1 are as shown in Fig. 5.4. The lag 1 autocorrelation for the e_{rj} is 0.38 with a standard error of 0.07, so that assumption of independence is questionable. In such cases, Zimmerman and Harville (1991) suggest fitting a correlation structure to the residual errors e.

An alternative approach, proposed by Gleeson and Cullis (1987), assumes that the spatial correlation is caused by the trend ξ which is modelled by sequentially fitting autoregressive–moving average (ARMA) or autoregressive–integrated–moving average (ARIMA) processes (Section 5.6). The errors e are then assumed i.i.d. $N(0,\sigma^2)$. Data-dependent diagnostic checks are used to assess the adequacy of the fitted model. We concentrate on the Gleeson–Cullis approach to spatial analysis, which incorporates most earlier 'neighbour' models proposed in the statistical literature for agricultural field trials (Wilkinson *et al.*, 1983; Green, Jennison and Seheult, 1985; Besag and Kempton, 1986; Williams, 1986a).

For Example 2, the Gleeson–Cullis approach led to an ARIMA (0,1,0)

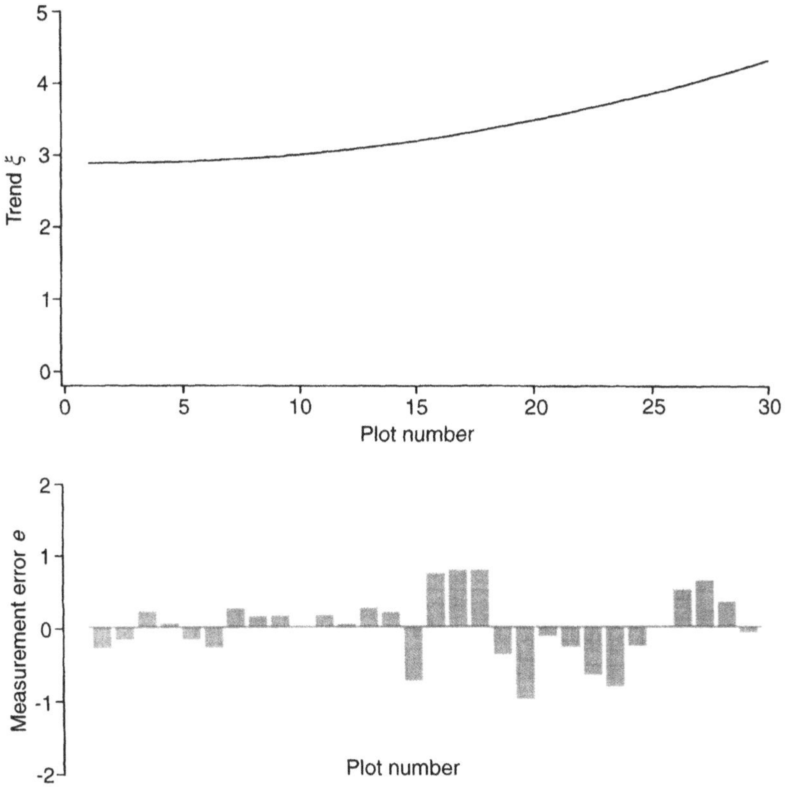

Fig. 5.4. Quadratic trend and errors for Example 2.

model for the ξ after assuming that the e values were i.i.d. $N(0,\sigma^2)$. For range 1, the estimates for trend and error are shown in Fig. 5.5. The lag 1 correlation for error e is 0.05 ± 0.18, so the assumption of i.i.d. $N(0,\sigma^2)$ errors appears adequate.

5.5 STATIONARITY

In spatial analysis, the trend ξ and error e components of the plot errors are assumed random. A fundamental question in modelling these random effects is whether or not they are stationary, i.e. whether the effects vary about a fixed mean (Wilkinson *et al.*, 1983). While techniques are available (e.g. Cressie, 1988) for checking stationarity, they are not widely used.

The plot errors from the complete block model for Example 2 (Fig. 5.2) do not appear to be stationary over the 30 plots of range 1. This is supported by the differences between sub-block means (Fig. 5.3), and, removing these means, the errors e from the incomplete block model appear far more stationary. In the Gleeson–Cullis spatial analysis approach, the trend ξ appears to be

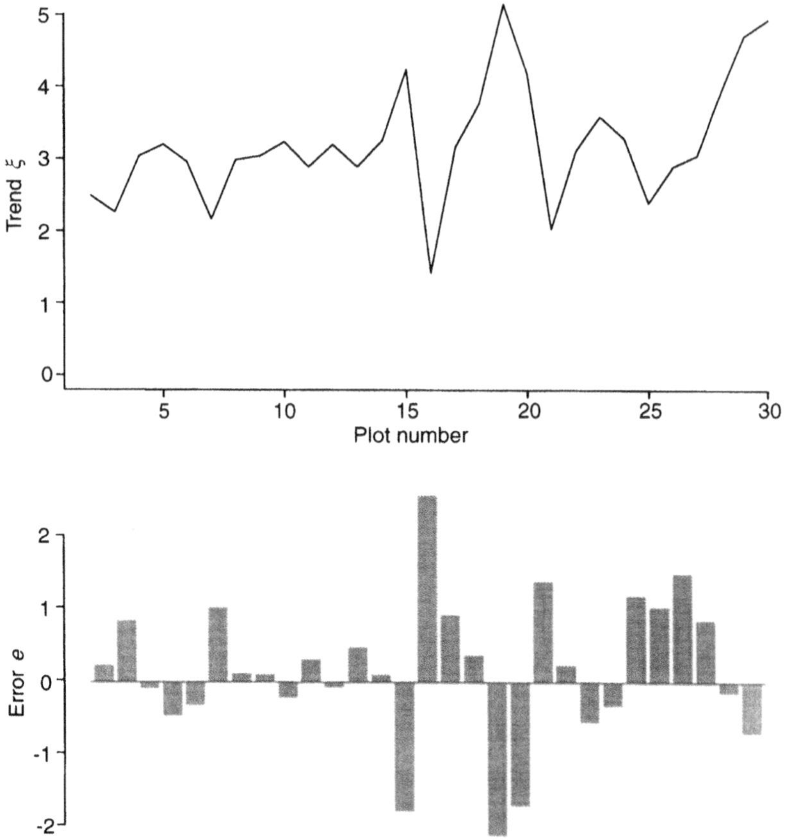

Fig. 5.5. Trend and errors from ARIMA (0,1,0) model for Example 2.

non-stationary whereas it seems reasonable to assume stationarity for the errors e (Fig. 5.5).

Non-stationarity in plot yields can often be removed by differencing the data. For Example 2, if we suspect non-stationarity of plot errors within each range, we can difference the data within the ranges to get a 6 × 29 data array:

$$
\begin{array}{ccccc}
y_{11} - y_{12} & y_{12} - y_{13} & y_{13} - y_{14} & \cdots & y_{1,29} - y_{1,30} \\
y_{21} - y_{22} & y_{22} - y_{23} & y_{23} - y_{24} & \cdots & y_{2,29} - y_{2,30} \\
. & . & . & \cdots & . \\
. & . & . & \cdots & . \\
. & . & . & \cdots & . \\
y_{61} - y_{62} & y_{62} - y_{63} & y_{63} - y_{64} & \cdots & y_{6,29} - y_{6,30}
\end{array}
$$

What effect does this have on the suspected non-stationary trend ξ? For range 1, we can graph the differenced trend effects $\xi_{11} - \xi_{12}$, $\xi_{12} - \xi_{13}$, ..., $\xi_{1,29} - \xi_{1,30}$, in plot order as shown in Fig. 5.6. The assumption of stationarity now appears much more tenable than it did for ξ.

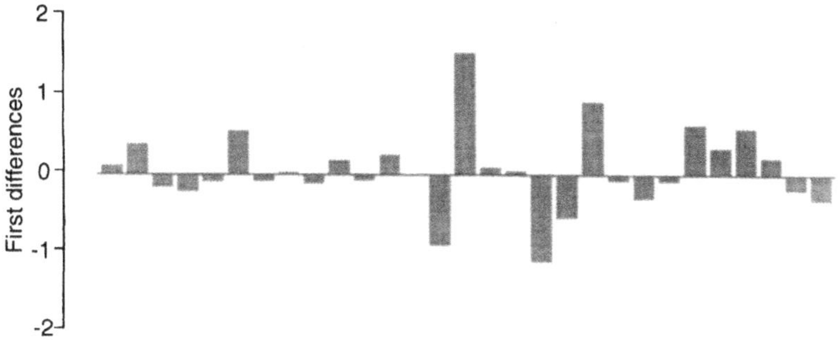

Fig. 5.6. Differenced trends from Fig. 5.5.

Occasionally, higher-order differencing may be required to achieve stationarity. Second-order differencing is equivalent to differencing twice. That is, we calculate

$$y_{11} - y_{12} - (y_{12} - y_{13}),\ y_{12} - y_{13} - (y_{13} - y_{14}), \ldots$$

to obtain

$$y_{11} - 2y_{12} + y_{13},\ y_{12} - 2y_{13} + y_{14}, \ldots$$

Differencing to order d (i.e. repeating the differencing process d times) will remove a trend of order $d - 1$. Thus first-order differencing will remove the range means (the block effects in Equation 5.2) and second-order differencing will remove a linear trend within each range. For the type of non-stationarity encountered in agricultural field trials, only low-order differencing (rarely greater than first) is needed to achieve stationarity.

Differencing a non-stationary process will eventually lead to a stationary process, whereas differencing a stationary process results in a stationary but more highly correlated process. We therefore advise that, provided the number of plots in a range is reasonably large, spatial analysis includes an initial differencing of data whenever there is doubt about the stationarity of the errors in a trial.

5.6 ONE-DIMENSIONAL SPATIAL ANALYSIS

One-dimensional spatial analysis is suitable for trials which exhibit the correlation pattern in the wheat uniformity trial of Example 2. The most frequently used spatial model is the 'first differences' model which is based on the assumption that plot variation has two components, the trend ξ and residual error e; the errors e are assumed i.i.d. $N(0,\sigma^2)$, and the first differences of trend ξ are also assumed i.i.d. $N(0,\sigma_t^2)$ where $\sigma_t^2 \neq \sigma^2$. The first differences of the trend are also assumed independent of the errors e. The 'first differences' model is a simple example of the class of ARIMA models which describe a wide variety of spatial patterns. We now briefly describe the class of ARIMA models.

Consider a series of plot yields $z_1, z_2, z_3, \dots$, which have associated errors $a_1, a_2, a_3, \dots$. An autoregressive process of order p, AR(p), arises when an observation is affected by p previous observations:

$$z_t = \phi_1 z_{t-1} + \phi_2 z_{t-2} + \dots + \phi_p z_{t-p}$$

The spatial correlation for an AR(p) process decreases with increasing lag. Figure 5.7 gives examples for an AR(2) process. The three examples of uniformity trials in Section 5.3 show a slow decay in correlation with increasing lag, suggesting that an autoregressive process with $\phi_1 > 0$ may be a suitable model.

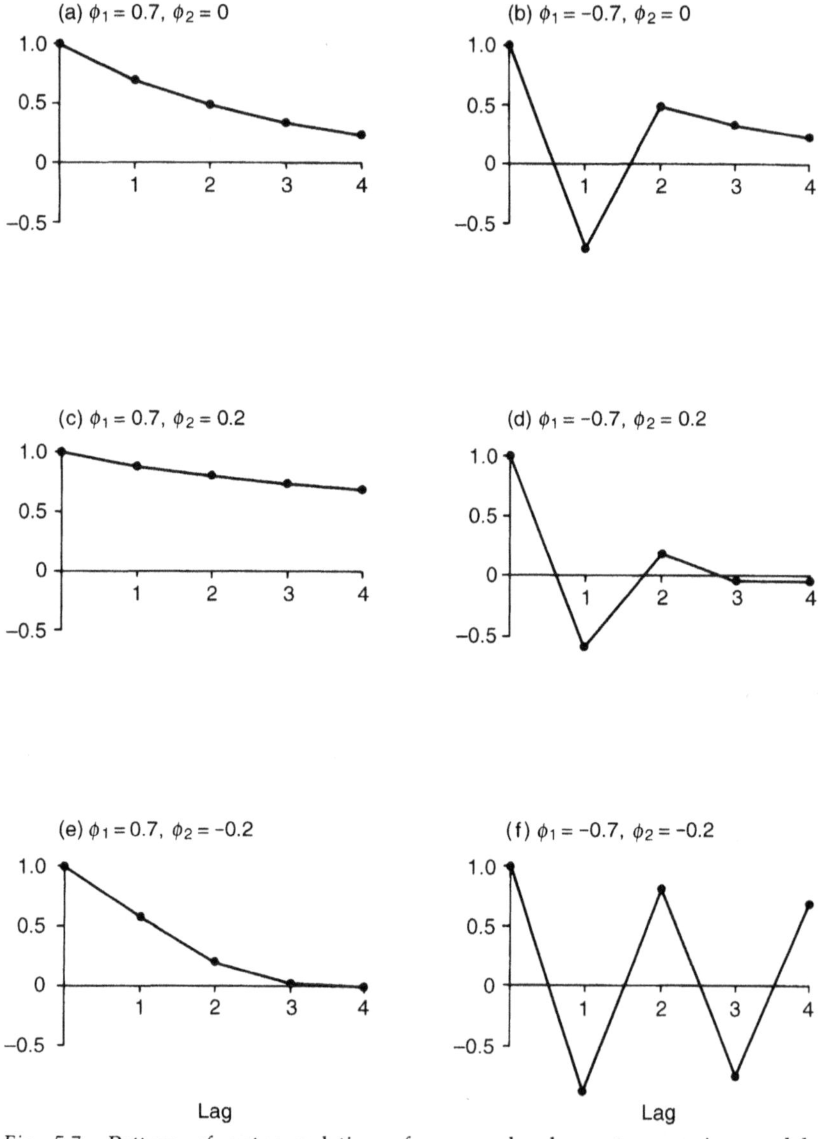

Fig. 5.7. Pattern of autocorrelations for second-order autoregressive model, $z_t = \phi_1 z_{t-1} + \phi_2 z_{t-2}$, for different parameter values, ϕ_1 and ϕ_2.

Alternatively, suppose an observation is affected by its plot error and errors from the preceding q plot errors. We then have a moving average process of order q, MA(q):

$$z_t = a_t - \theta_1 a_{t-1} - \theta_2 a_{t-2} - \dots - \theta_q a_{t-q}$$

The spatial correlations for an MA(q) process have a characteristic pattern, in which only the first q lag correlations are non-zero (Fig. 5.8). Differencing a series of observations which have a slow decay in the correlations often produces such a correlation pattern.

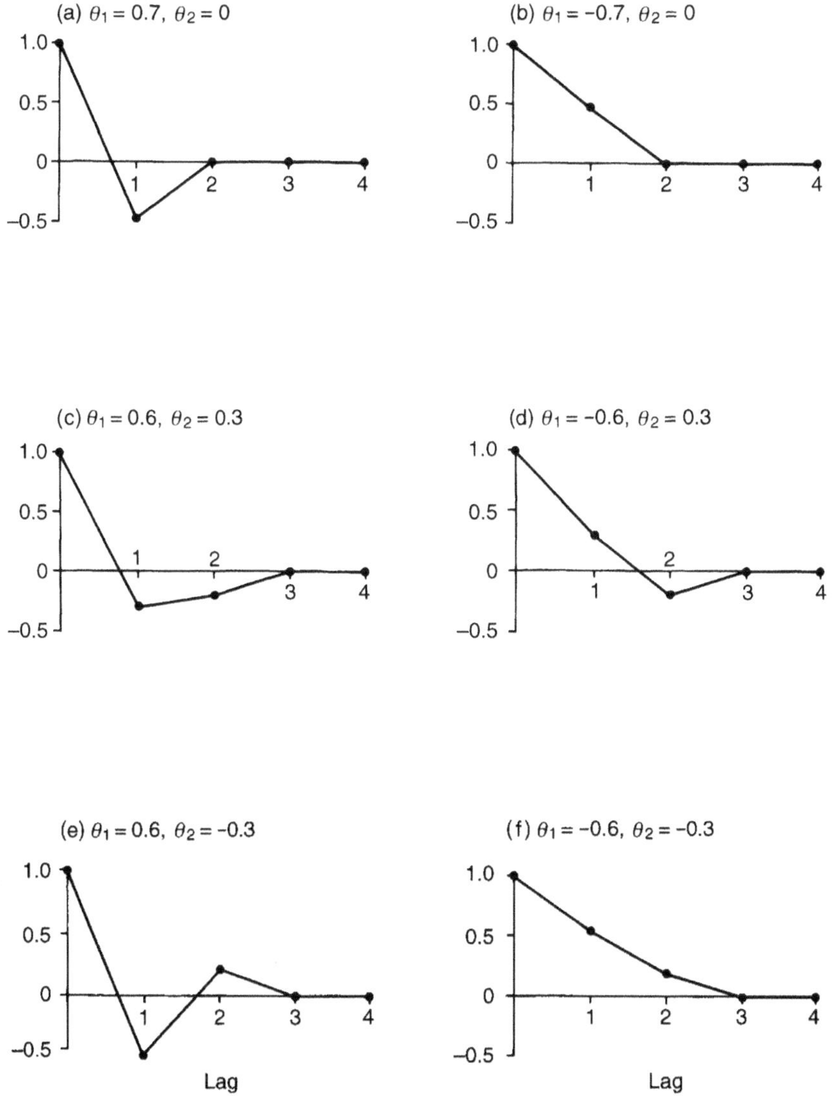

Fig. 5.8. Pattern of autocorrelations for the second-order moving average model, $z_t = a_t - \theta_1 a_{t-1} - \theta_2 a_{t-2}$, for different parameter values, θ_1 and θ_2.

In practice, the fewer parameters we have to estimate to describe a correlation pattern, the better; a principle known as parsimony. It is rarely necessary to choose p or q to be larger than one for agricultural field trials. However, both autoregressive and moving average terms may be needed: the most frequently used model is then ARMA (1,1), the autoregressive (order 1)–moving average (order 1) model.

ARMA models are applicable only to stationary processes which are not always appropriate for field trial data. In Section 5.5, however, we showed how differencing of plot data can lead to stationarity. A further advantage of differencing is that it often simplifies the correlation of pattern and so leads to a more parsimonious model.

To reconstruct the original data from a process which has been differenced say d times, we need to sum, or integrate, the process d times. If the differenced data follow an ARMA (p,q) model, the original data follow an ARIMA process (p,d,q). In agricultural field trials first differencing is usually sufficient.

If we assume the plot errors have two components, a trend ξ, assumed to be an ARIMA (p,d,q), and a residual error e, assumed to be i.i.d. $N(0,\sigma^2)$, then the differenced plot errors ($\xi + e$) can be represented by an ARMA (p,Q) where Q is the maximum of $p + d$ and q (Box and Jenkins, 1970). For the 'first differences' model, where the trends ξ are assumed to follow an ARIMA (0,1,0), and the errors e are assumed i.i.d. $N(0,\sigma^2)$, the first differences of errors can be directly fitted by an ARMA (0,1) or first-order moving average MA(1). Hence the 'trend plus error' model is ARIMA (0,1,1).

5.7 ESTIMATION

The main purpose of spatial analysis is to estimate genotype effects and their standard error of differences. However, these will depend on the choice of model and the values of variance parameters for trend and error. These parameters are estimated by residual maximum likelihood (REML) and used as weights for generalized least squares estimates of the genotype effects (Gleeson and Cullis, 1987).

Lill, Gleeson and Cullis (1988) assessed the accuracy and efficiency of REML estimation for one-dimensional spatial analysis in a large simulation study. REML estimates of variance parameters were shown to be very accurate and resulting estimates of genotype effects showed substantially increased precision compared with estimates from an incomplete block analysis. The improvement in precision increased as the within-trial heterogeneity increased.

In a spatial analysis, there is no assumption of the independence of plot yields; hence the estimates of genotype effects are themselves not independent. The analysis thus provides a separate SED for each pairwise comparison (as for incomplete blocks), though with good design the SEDs are similar and, for convenience, the average SED is often used as a summary statistic.

5.8 ASSESSING MODEL ADEQUACY

Gleeson and Cullis (1987) suggested a sequence of ARIMA models be fitted to the data, and diagnostic checks used to assess whether a particular model adequately represents the observed pattern of variation. Formal statistics, which can be used as diagnostics for checking model adequacy, include likelihood ratio tests, partial autocorrelations (Box and Jenkins, 1970), *Q*-tests based on the sums of squares of residual autocorrelations (Ljung, 1986) and autocorrelations of recursive residuals (Cullis, McGilchrist and Gleeson, 1991).

The Gleeson–Cullis approach uses information from several tests, in conjunction with other diagnostics such as plots or sample correlations of residuals. A three-step strategy has been found useful.

1. Examine the graphs of residuals plotted in field order – this will show up non-uniform patterns in the residuals (rather than uniform distribution, as desired) and highlight potential outliers (see Chapter 3).
2. Examine the sample correlations of 'whitened' residuals (Gilmour, 1992). Based on the results of Cullis, McGilchrist and Gleeson (1991), we can derive an approximate *t*-test for the lag 1, lag 2, ... autocorrelations of the 'whitened' residuals for the fitted model. This test applies only to sample autocorrelations along the principal axes of the sample correlation matrix (i.e. for the same row or column), but this usually suffices even for the two-dimensional spatial model considered in Section 5.9.
3. The likelihood ratio statistic can be used to assess the order of the model to be fitted, i.e. to test whether an MA(2) model is a better fit than an MA(1), or an ARMA (1,1) is better than an AR(1). However, its use is limited to those models fitted under the same regime of differencing.

5.9 EXTENSION TO TWO-DIMENSIONAL SPATIAL ANALYSIS

Modelling spatial variation in one dimension often suffices in trials with long narrow plots, or where the small number of ranges make modelling the spatial variation between ranges impractical.

However, the level of variation in both dimensions (between ranges and between passes) sometimes warrants a two-dimensional approach. Robinson, Kershaw and Ellis (1988) examined variation in barley breeders' small plot trials in the UK, and concluded that most trials would show gains in efficiency from two-dimensional analysis. Cullis and Gleeson (1991) examined 24 uniformity trials covering a range of crops and found that in half the trials a two-dimensional analysis was warranted.

Martin (1990) introduced separable ARMA processes for the two-dimensional spatial analysis of agricultural field trials. Cullis and Gleeson (1991) adopted a similar approach using separable ARIMA processes. Zimmerman and Harville (1991) proposed modelling the large-scale variation in two dimensions by polynomial trends, then modelling the small-scale variation in both dimensions via a stationary covariance structure. This is equivalent to fitting separable ARIMA processes.

What is meant by separable? If we have sample spatial correlations $\bar{r}(u,v)$ then for a separable process:

$$\bar{r}(u,v) = r(u,0) \times r(0,v)$$

that is, the overall covariance matrix can be expressed as a direct product of the covariances for the two principal directions. The assumption of separability greatly reduces the computation necessary for fitting two-dimensional models to data as the matrices in the iterative estimation procedure are of considerably smaller size.

An additional benefit of separability is that error model diagnostics can be applied to the residual spatial correlations along the principal axes, i.e. first to the model fitted in one dimension, then to the model fitted to the second dimension. A sensible strategy for model choice is to start with the simplest models and fit progressively more complex models until an adequate fit is obtained. The approach we have found useful in analysing plant breeders' trials in the New South Wales Department of Agriculture is as follows.

1. For a trial where the size of the array of plots is sufficiently large in both directions (i.e. at least eight ranges and at least eight plots per range). Examine the spatial correlation matrix of the treatment-corrected data and, if there is substantial correlation in either or both directions, difference the data in the direction of the greater correlation and fit a first-order moving average model. Fitting a null correlation (independent) model to differenced data assumes that all the spatial variation in that direction is a trend, and there is no measurement error. This is generally unrealistic, so we recommend fitting at least an MA(1) to first differenced data. Examine the spatial correlation of the whitened residuals after fitting the MA(1) model to see if it adequately removes the correlation in that direction; if not, fit a higher-order model. When an adequate fit is obtained in the first direction, examine the correlation in the other direction (remember that differencing in one direction can alter the correlation pattern in the other direction). If substantial correlation remains in the second direction, first difference the data in the second direction and fit an MA(1), or successively higher order models until the spatial correlation matrix of the whitened residuals is satisfactory.
2. If the array of plots is small in one direction (< 8). Differencing is rarely warranted for a short series and where a range includes an incomplete set of genotypes, it will result in a loss of information on genotypes from the range means (Section 4.7; Kempton, Seraphin and Sword, 1994). Fitting an autoregressive process to the undifferenced data may then be a better option. Again, first examine the spatial correlation matrix of the treatment-corrected data. If one of the dimensions is sufficiently large (≥ 8) and there is substantial correlation in that direction, first difference and fit an MA(1) or higher-order model as before. We recommend treating this longer direction first even if the observed correlation is larger in the shorter direction; the correlation of the whitened residuals in the shorter direction is likely to be reduced by differencing in the longer direction. Having obtained an adequate model for the longer direction, examine the spatial correlations

in the shorter direction and try fitting autoregressive models to describe that pattern, beginning with a first-order autoregressive model or AR(1).

3. Where the size of the array is small (< 8) in both dimensions, but the spatial correlations are substantial, try fitting autoregressive or moving average models to the undifferenced data, first in the direction of greater correlation then, if necessary, the other direction.

This strategy provides a rough guideline for the experimenter; when there are anomalies in the data that are hard to explain, consult an experienced statistician for assistance.

5.10 EXAMPLE

Table 5.5 gives the field layout and plot yields from a breeder's trial to compare yields of 25 spring wheat varieties conducted at Slate Hall Farm, Cambridgeshire, UK, in 1976. The plot size was 1.5 m × 4.5 m and the experimental design was six replicates of a balanced (5 × 5) lattice square in a 10 × 15 array.

If we assume that all of the plots are independent, the average SED between the genotype means is 121. The spatial correlation matrix of the data for this model (i.e. corrected for genotype means) is given in Table 5.6(a) and the fitted model is clearly inadequate.

Following the strategy outlined in the Section 5.9 we difference the plot yields within the ranges and fit an MA(1) to the between plots variance. After fitting this model on ARIMA (0,0,0) × ARIMA (0,1,1) the average SED of genotype means reduces to 67. The corresponding spatial correlations are given in Table 5.6(b). The spatial correlations of the whitened residuals along the principal axis (between plots within a range) is now consistent with the assumptions of independence but there is still substantial correlation between ranges. We now difference in the second dimension and fit an MA(1) which further reduces the average pairwise SED of genotype means to 61. The spatial correlation matrix of the whitened residuals is presented in Table 5.6(c) and indicates that an ARIMA (0,1,1) × ARIMA (0,1,1) is an adequate fit to the variation pattern of this experiment.

5.11 CONCLUSION

We now return to our opening question: why spatial analysis? We have suggested why spatial models may be applicable to field experiments, but without appropriate software, a good working knowledge of the models available and tests of model adequacy, a spatial analysis will be more difficult to apply than a conventional randomization-based analysis. The answer lies in the potential of spatial analysis for improving estimates of genotype effects and their errors.

Several investigators have demonstrated the greater efficiency of spatial analysis. Patterson and Hunter (1983), in a study of cereal variety trails in the

Table 5.5. Genotype code (italics) and plot yields (g) of spring wheat at Slate Hall Farm, Cambridgeshire, in 1976. The trial consisted of six replicates of 25 genotypes laid out as 10 ranges of 15 1.5 m × 4 m plots

1/1003	*2*/1356	*4*/1412	*3*/1239	*5*/1508	*19*/1967	*23*/1572	*2*/1969	*6*/1747	*15*/1598	*18*/1630	*25*/1633	*9*/1255	*11*/1277	*2*/1572
6/1531	*7*/1540	*9*/1250	*8*/1658	*10*/1185	*8*/1605	*12*/1550	*16*/1500	*25*/1642	*4*/1504	*5*/1680	*7*/1526	*16*/1452	*23*/1480	*14*/1482
21/1126	*22*/1400	*24*/1329	*23*/1287	*25*/1555	*11*/1395	*20*/1696	*24*/1570	*3*/1404	*7*/1285	*6*/1473	*13*/1761	*22*/1695	*4*/1364	*20*/1790
11/1261	*12*/1423	*14*/1110	*13*/1735	*15*/1617	*22*/1820	*1*/1351	*10*/1297	*14*/1412	*18*/1506	*24*/1512	*1*/1355	*15*/1524	*17*/1478	*8*/1371
16/1458	*17*/2036	*19*/2119	*18*/1912	*20*/1893	*5*/1748	*9*/1450	*13*/1740	*17*/1450	*21*/1523	*12*/1364	*19*/1690	*3*/1334	*10*/1239	*21*/1557
3/1623	*18*/1862	*8*/1645	*13*/1888	*23*/1527	*16*/1606	*24*/1842	*10*/1186	*13*/1462	*2*/1242	*10*/1082	*4*/1304	*17*/1267	*11*/1266	*23*/1200
1/1331	*16*/1417	*6*/1611	*11*/1454	*21*/1790	*12*/1767	*20*/1917	*1*/1264	*9*/1060	*23*/951	*12*/1130	*6*/1266	*24*/1289	*18*/1260	*5*/1174
5/1211	*20*/1411	*10*/1183	*15*/1550	*25*/1660	*4*/1526	*7*/1681	*18*/1545	*21*/1290	*15*/976	*19*/1240	*13*/1181	*1*/917	*25*/1287	*7*/975
2/1388	*17*/1453	*7*/1384	*12*/1669	*22*/1738	*25*/1845	*3*/1700	*14*/1528	*17*/1373	*6*/1240	*21*/1252	*20*/1591	*8*/1428	*2*/1509	*14*/1273
4/1443	*19*/1667	*9*/1549	*14*/1459	*24*/1722	*8*/1583	*11*/1490	*22*/1607	*5*/1315	*19*/1174	*3*/1443	*22*/1649	*15*/1407	*9*/1315	*16*/1318

Table 5.6. Spatial correlations of plot yields for the Slate Hall Farm trial after correction for genotype effects: (a) null model; (b) ARIMA (0,0,0) × ARIMA (0,1,1) model; (c) ARIMA (0,1,1) × ARIMA (0,1,1) model

Distance apart (ranges)	*Distance apart (plots within a range)*			
	0	*1*	*2*	*3*
(a) 0	1.00	0.64	0.43	0.21
1	0.46	0.41	0.28	0.13
2	0.26	0.19	0.15	0.05
(b) 0	1.00	0.08	0.03	−0.14
1	0.29	−0.19	0.03	−0.07
2	0.27	0.08	0.07	−0.05
(c) 0	1.00	0.04	0.05	−0.03
1	0.06	0.10	0.03	0.02
2	0.03	−0.09	0.05	−0.00

UK, found a 44% reduction in average pairwise variance of variety contrasts using spatial analysis rather than randomized complete block analysis, compared with a 30% reduction from incomplete block analysis. Similarly, Gleeson and Cullis (1987), in a study of cereal variety trials in Australia, found a 42% reduction in average pairwise variance using spatial analysis, compared with a 33% reduction from incomplete block analysis. So with increased access to better software, maybe the question should now be: why not spatial analysis?

A computer package TwoD (Gilmour, 1992) for two-dimensional spatial analysis of agricultural field trials is available for minicomputers, UNIX workstations and PCS. TwoD is currently being incorporated into the general purpose statistical package, GENSTAT.

6
Unreplicated trials

R.A. KEMPTON AND A.C. GLEESON

6.1 INTRODUCTION

Chapters 4 and 5 present methods of design and analysis for replicated trials. Replication increases the precision of estimates of genotype differences and also provides a measure of that precision, which is important for predicting the yields of genotypes immediately prior to commercial release. In the early generations of a breeding programme, however, the benefits from replication are less clear as the main interest is on ranking genotypes rather than predicting their yields. Bos (1983a) showed theoretically that, with fixed resources, the expected genetic gain may be greater for a selection based on a broad screening of many genotypes, rather than on a more precise assessment of a smaller number (see Section 1.2). Furthermore, with multilocation testing, the commitment of large resources to maximizing the precision of comparisons among genotypes within individual locations may be wasteful because genotype × environment variation limits the precision of across-location comparisons (Chapter 10). Replication may also be limited by seed supplies in the early generations.

In this chapter, we consider methods of design and analysis for controlling the local effects of field variation (local control) in unreplicated trials. Despite its long history in plant breeding, the topic is rarely considered in textbooks on experimental design and analysis: Roemer (1930, pp. 135–155) gives a detailed account of the methods used early this century while Kempton (1984a) provides a more recent review.

We consider three methods of local control: (i) comparing genotypes with local checks; (ii) comparing genotypes with neighbouring genotypes; (iii) fitting a spatial model following the approach of Chapter 5.

6.2 LOCAL CONTROL USING CHECK PLOTS

Check plots were used extensively for controlling spatial heterogeneity, particularly in the USA (Thorne, 1907; Wiancko, 1914; Pritchard, 1916; Stadler,

Statistical Methods for Plant Variety Evaluation. Edited by R.A. Kempton and P.N. Fox. Published in 1997 by Chapman & Hall, London. ISBN 0 412 54750 3.

1921), before the benefits of replication and blocking were generally established. Replicated checks, often of established varieties, measure the yield variation across the trial, and the yields of unreplicated test genotypes are then assessed against local checks. Holtsmark and Larsen (1905) give an example of a design in which a single check variety is sown in every third plot and test genotypes are compared with the mean of three neighbouring checks (Fig. 6.1). We call the mean yield of the checks used to adjust genotype yields, the environmental index (denoted by z). A major advantage of using checks is that genotypes need not be randomized across the trial. Visual evaluation in the field may be simplified by grouping genotypes, for example by parentage, and using a regularly spaced grid of checks.

g	g	**A**	g	g	A
g	**A**	g_0	g	A	g
A	g	g	**A**	g	g
g	g	A	g	g	A

Fig. 6.1. Design for unreplicated genotypes g with a check A in every third plot (Holtsmark and Larsen, 1905). The yield of genotype g_0 is adjusted for field variation by subtracting the mean yield of the three neighbouring checks (in bold type).

6.2.1 Choice of check varieties

Local control is only effective if the yields of the checks show the same general pattern of response across the trial as the test genotypes. Otherwise, adjustment of test plot yields by yields of the local checks may increase, rather than reduce, the error of assessment (Salmon, 1914). This could occur, for example, if the check variety was more susceptible than the test genotypes to a soil-borne pathogen which varied in prevalence across the trial. In practice, it is advisable to use at least two check varieties with a similar genetic background to the test material. Stadler (1921) conducted a replicated oat trial in which each genotype was grown between plots of two check varieties. He found that the effectiveness of the adjustment to genotype yields depended on which check variety was used for local control. The replicate to replicate variation of genotype yields was usually smallest when the yields of progeny were adjusted using a parent as the check.

6.2.2 Frequency of checks

In general, increasing the frequency of check plots will enhance local control and hence the genetic gain from selection, provided the selection intensity is kept constant. However, when the trial size and number of genotypes selected at a given stage are fixed, inclusion of check plots will reduce the number of genotypes tested and, concomitantly, the selection intensity. Thus the checks must increase genotypic precision just to maintain genetic gain. This is illustrated in Fig. 6.2 where the number of selected genotypes is set at one tenth of the total number of plots in the trial. When the frequency of checks is high, the plot error must be reduced substantially to improve the efficiency of selection, particularly when the heritability h^2 of the genotypes in the trial is high. Indeed, when $h^2 > 0.64$, using checks in alternate plots (a frequency of one in two) will always result in a smaller expected genetic gain than selecting from twice the number of genotypes with no checks. In general, when the total number of plots is fixed, using a high frequency of checks (say greater than one in five) is unwise unless the spatial heterogeneity is large and very local so that correlations among plot yields fall off rapidly with separation.

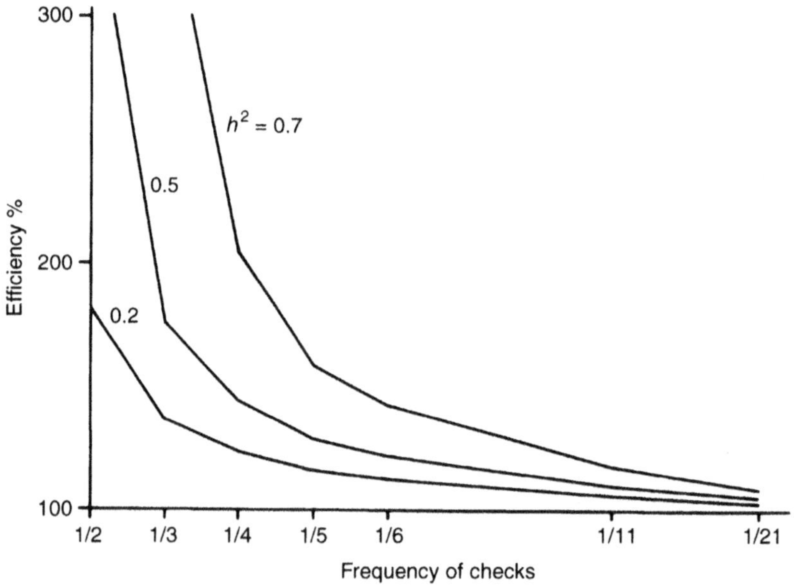

Fig. 6.2. Efficiency of check plot adjustment required to maintain genetic gain from the selection for different frequencies of check plots and different base-line heritabilities h^2 when the total number of trial plots n and number of selected genotypes $n/10$ are fixed.

6.2.3 Arrangement of checks and method of adjustment

The simplest experimental designs are based on a systematic arrangement of equally spaced checks. When the plots are long and narrow, genotype yields may be adjusted by an environmental index constructed from the yields of checks in the same range. Four commonly used indices are:

1. the yield of the nearest check;
2. the mean yield of the two nearest checks, one on either side of the test plot;
3. the weighted mean of these two checks, where weights are inversely related to the distance from the test plot; and
4. the mean yield of all the checks in the range.

If several check varieties are used, the check plot yields are first adjusted for variety effects. When there are substantial fertility trends along ranges so that the correlation between the plots decreases rapidly with separation, method 3 will generally give the most efficient adjustment (Kempton, 1984a); otherwise, method 4 is superior.

More complex augmented designs were considered by Federer in papers spanning 20 years (see Federer and Raghavarao, 1975). In the simplest designs (Federer, 1961), a number of check varieties are laid out in a complete or incomplete block design and each block is augmented with unreplicated genotypes (Section 4.5.6). A standard analysis of the check plot yields gives estimates of the block and variety effects, and each test genotype is then adjusted for the block in which it occurs. Federer (1961) gives a method for calculating the variances of differences in means between two genotypes, and between a genotype and a check variety. The designs have the disadvantage that the positions of checks must be randomized within each block, producing an irregular field layout which complicates visual assessment when compared with a systematic design. The designs are also impracticable for large numbers of genotypes as they require large numbers of check plots to control variation.

Augmented forms of row and column designs have also been developed to allow genotype yields to be adjusted for field variation in two dimensions (Federer and Raghavarao, 1975; Lin and Poushinsky, 1983). The design in Fig. 6.3 is based on a 4×4 latin square for four check varieties using square plots. Each check plot is the centre of a 3×3 block of plots and is surrounded by unreplicated genotypes, so that the frequency of checks to test genotypes is one in eight. Genotype yields are adjusted for block effects which may be derived as follows. An analysis of variance of check plot yields gives estimates of row, column and variety effects, and block effects are estimated as the sum of row and column effects or, if there is strong local heterogeneity, from central check plot yields corrected for variety effects. Lin and Poushinsky (1983) suggest including further checks in some test plots to investigate the correlation between the outer and centre plots within blocks, and so refine the adjustment.

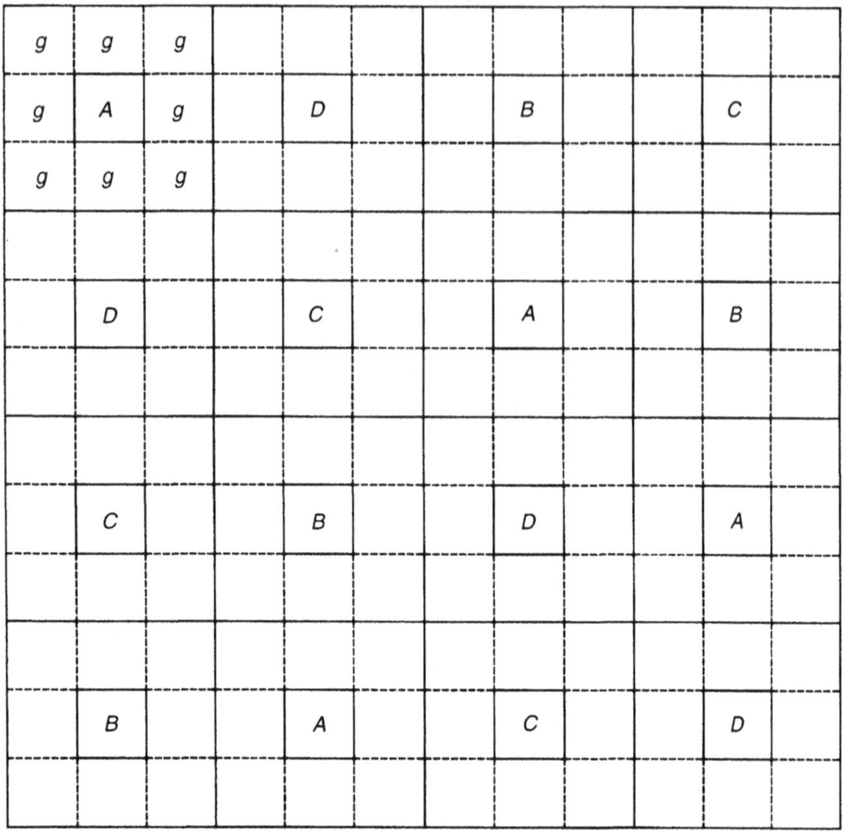

Fig. 6.3. Augmented latin square design based on four checks *A*, *B*, *C* and *D* with 144 unreplicated test genotypes *g* in square plots (Lin and Poushinsky, 1983).

6.2.4 Validity of adjustment by check plots

We noted in Section 6.2.1 that when checks show a different environmental response from test genotypes, adjustment by checks will increase rather than decrease the error of genotype comparisons. To investigate the pattern of adjustment, we recommend that check plot yields are displayed as a smoothed contour map. This will show up major fertility trends across the trial area and indicate when a correction of genotype yields might be effective. When more than one check variety is used, separate maps can be produced for each variety and checked for consistency (Section 6.2.5).

When genotypes are replicated, the risk of inappropriate adjustment is reduced by using the environmental index z as a covariate in the analysis of variance (Richey, 1924; Yates, 1936). Kempton and Talbot (1988) suggested that a small number (say 5%) of test genotypes be duplicated and an analysis of covariance carried out on the yields of these genotypes to estimate the regression coefficient β for the covariate z. All genotypes are then adjusted by

βz. If the covariance analysis indicates that β is not significantly different from zero, the adjustment is unlikely to improve selection and may be omitted.

Inter-plot competition may mean that a check is a poor yardstick for assessing genotypes (Chapter 7). Jensen and Federer (1964) analysed plot yields of winter wheat trials, where test genotypes were grown in three-row plots with a single-row plot of a check after every tenth plot. The test material included tall and semi-dwarf genotypes, with a tall variety used as a check. The yield of the check was found to be negatively correlated with the height of its adjacent plot and thus ineffective as an environmental indicator. Indeed, if the check were used for adjustment, there would be a bias towards selecting for taller genotypes (Fischer, 1979). Consequently, Jensen and Federer (1964) recommended three-row check plots for their winter wheat trials with only the centre row of each plot harvested for yield.

6.2.5 Example of check plot adjustment for a winter wheat selection trial

Besag and Kempton (1986) analysed a selection trial of F_5 generation winter wheat carried out at the former Plant Breeding Institute, Cambridge, UK. A total of 1560 test genotypes were grown in unreplicated plots, approximately 1.5 m wide by 4.5 m long. Check plots of two alternating varieties, 'Maris Huntsman' and 'Bounty', occupied every sixth plot [Fig. 6.4(a)]. The trial consisted of 52 ranges of 37 plots and exceeded two hectares.

Examination of the check plot yields showed that both varieties had a two-fold variation in yield over the trial. Although 'Bounty' out yielded 'Maris Huntsman' by 9% on average, their smoothed contour maps for yield show a consistent pattern of environmental response (Fig. 6.5). As inter-plot correlation decreased only slowly with separation, particularly within ranges, Besag and Kempton (1986) constructed an environmental index for each plot based on the weighted mean yield of six closest checks, with weights as shown in Fig. 6.4(b). The selection of test genotypes was based on the plot yields adjusted for the environmental index.

(a)
A g g g g g B g g g g g A . . . . .
B g g g g g A g g g g g B . . . . .
A g g g g g B g g g g g A . . . . .
B g g g g g A g g g g g B . . . . .
A g g g g g B g g g g g A . . . . .
. . . .

Direction of sowing

(b)
⅛ A B ⅛
¼ B g g g g g A ¼
⅛ A B ⅛

Fig. 6.4. (a) Arrangement of two check varieties, *A* and *B*, and (b) weights used to construct the environmental index for assessing unreplicated test genotypes, *g*.

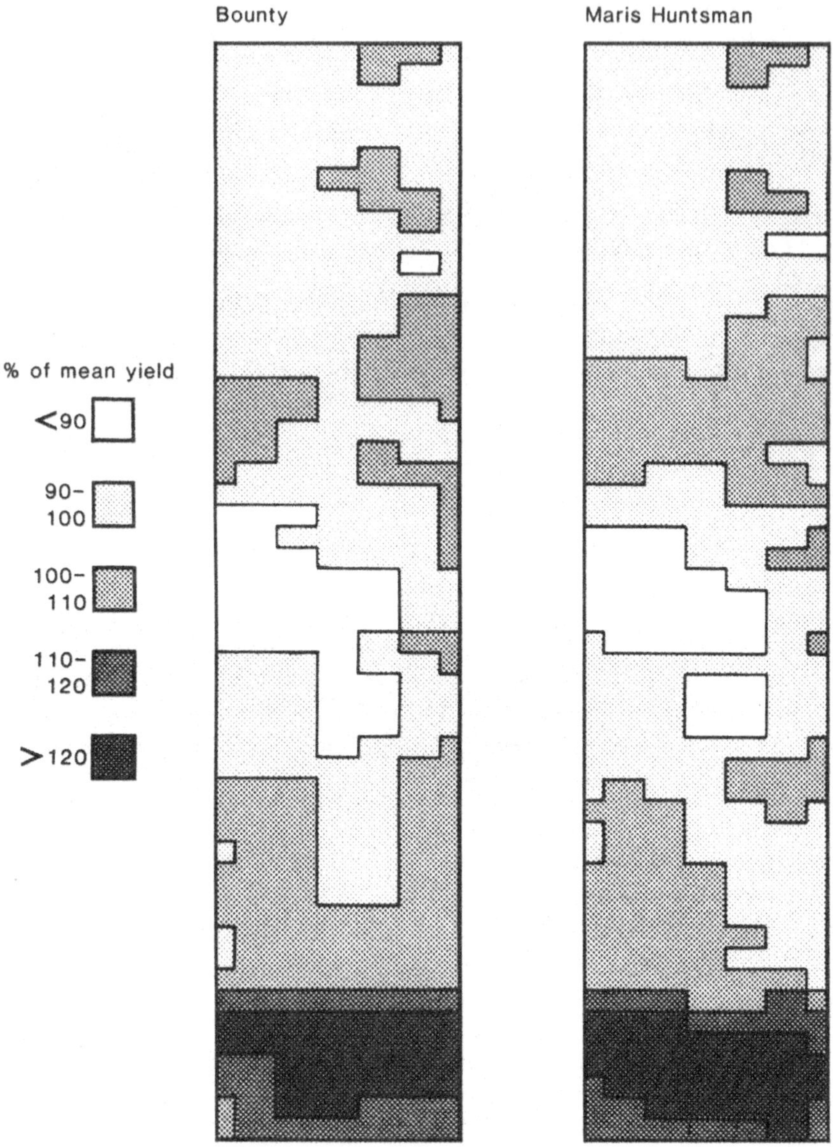

Fig. 6.5. Fertility map of the trial area based on the moving average of the check plot yields for two winter wheat varieties, 'Maris Huntsman' and 'Bounty' (Besag and Kempton, 1986).

The effect of adjustment on selection is illustrated in Fig. 6.6 which shows the field location of the 10% highest yielding test genotypes. When unadjusted yields are used, the selected genotypes come almost wholly from the area of the field with the highest fertility, but, after adjustment, they are

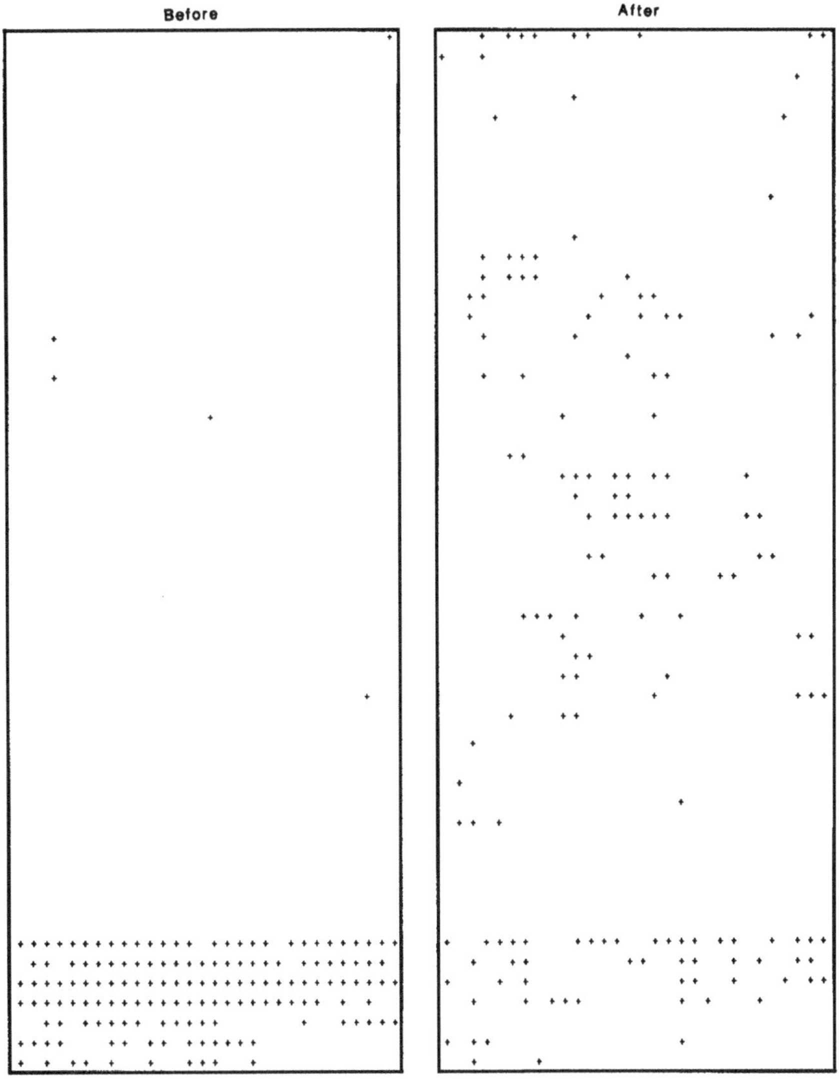

Fig. 6.6. Field location of the 10% highest yielding genotypes from a selection trial of 1560 winter wheat genotypes, before and after adjustment for an environmental index calculated from the check plot yields (Besag and Kempton, 1986).

widely distributed across the trial. Selections still show some clustering, however, as genotypes from the same cross were sown together.

6.3 LOCAL CONTROL FROM BLOCKING TEST PLOTS

An alternative method of controlling local fertility trends is to assess each test genotype in relation to neighbouring genotypes rather than independent checks. For this to be effective, genotypes should be fully randomized across the trial.

A straightforward method of selection divides the trial area into blocks and selects the highest yielding genotype within each block. This has been termed grid selection (Bos, 1983b) and it is also used for single plant selection (Gardiner, 1961). Choice of block size (k plots) and number of genotypes selected per block (n) fixes the selection proportion ($p = n/k$). In a variant of grid selection, the blocks are not fixed but move to the centre on each genotype in turn. Choosing, for example, blocks of 3×3 plots, a genotype is selected if it outyields the eight adjacent plots. With single plant selection, a wide range of block sizes and shapes may be used (Stam, 1984). For example, Fig. 6.7 shows the selections based on comparisons with the six adjacent plants when plants are sown in a honeycomb pattern (Fasoulas, 1973). The moving grid approach may lead to more effective local control but has the disadvantage that the selection proportion cannot be specified by the breeder in advance. A further restriction relates to the number of genotypes selected from each block: for a fixed grid selection, n is the same for each block and specified in advance; for the moving grid, n is zero or one.

34	30	**69**	23	48	28	14	51	34	38	30	49	**57**	
27	**56**	41	32	46	48	25	**72**	23	29	**40**	26	26	27
48	14	28	51	**71**	★	48	29	27	31	39	33	24	
32	38	20	18	52	19	25	20	**43**	37	**44**	31	14	35
50	43	44	**67**	18	17	44	36	33	28	26	8	32	
86	59	45	28	19	46	**54**	11	**46**	41	18	26	30	**32**
54	46	45	**49**	42	36	35	27	15	35	44	30	26	
45	**88**	44	33	46	36	15	27	38	35	47	**49**	28	24
84	71	42	27	**80**	72	38	33	48	**52**	46	34	38	
25	56	56	16	40	7	23	33	32	34	30	39	32	27
65	68	38	21	32	38	44	35	**55**	34	22	18	24	
50	**80**	13	44	12	52	**65**	30	22	20	42	29	**58**	54

Fig. 6.7. Yields (g) of single wheat plants from part of a selection trial with plants arranged in a honeycomb pattern (Fasoulas, 1973). Bold numbers indicate the plants selected on basis of superiority over their six nearest neighbours. The empirical selection proportion is 0.15.

Another approach, based on fixed blocks, is to adjust each plot yield by the mean of all plots in the same block and select on a whole trial basis using these adjusted values (Schutz and Cockerham, 1966; Bos, 1983b). The moving block analogue is the widely used method of moving averages (Richey, 1926; Stam, 1984). Using the means of test genotypes rather than independent checks to derive the environmental index is problematic. When spatial variation is fine grained, the blocks should be small and will then contain few genotypes, but the block mean may then provide a poor environmental index because of confounding with random genotype effects. The best block size depends on the ratio of genetic to environmental variance: where this ratio is high, large blocks should be chosen, and vice versa. The method also requires that the yields of all trial plots are measured, so poor genotypes cannot be discarded prior to harvest, as is possible using grid selection or check plots. This problem may be circumvented by adjusting genotypes using the median yield in each block, rather than the mean (Kempton, 1984a). For single plant selection, Bos (1983b) suggested standardizing the adjusted plot values by dividing by the within-block standard deviation before selection. Otherwise, he argues, more genotypes will tend to be selected from those parts of the field with the greatest local variation in yield.

The moving average method may be extended to using a weighted mean of neighbouring plots for adjustment, where the weights are inversely related to the distance from the central plot. Stam (1984) gives best linear unbiased estimators (BLUEs) for the weights when the spatial correlations among plots are known. He also compared the precision of moving average and BLUE estimates of genotype effects in a simulation study involving different ratios of genetic to environmental variance, scales of spatial variation and numbers of neighbours (block size). He concluded that BLUEs provide little improvement over a moving average adjustment with equal weights, once a neighbourhood size appropriate to the level of genetic and environmental variation is chosen.

Kempton (1984a) compared the efficiency of adjustment using a moving average of test plots with that using check plots. With a fixed total number of plots, he found that a moving average adjustment gave the higher expected genetic gain. Nevertheless, check plots are still commonly used to aid visual assessment in the field, so a combined analysis using local check varieties and test genotypes would be beneficial. Such an approach is described in the next section.

6.4 LOCAL CONTROL BY SPATIAL ANALYSIS

We now show how the formal methods of spatial analysis, introduced in Chapter 5, can be used to combine information from neighbouring plots, and possibly other covariates, in order to adjust plot values in unreplicated trials. It is assumed that genotypes are fully randomized across the trial.

6.4.1 Model and estimation

Cullis *et al.* (1989) extended the spatial analysis of the replicated field experiments to trials with only one plot of each test genotype but several plots of each check variety.

They defined an additive model for the yields of all plots as

$$\text{plot yield} = \text{genotype effect} + \text{covariate effects} + \text{trend} + \text{error} \quad (6.1)$$

where genotype effects correspond to both test genotypes and checks. The effects for check varieties are assumed to be fixed (as in replicated experiments), whereas effects for test genotypes come from a Normal distribution with mean μ and variance σ_g^2. The model can include additional covariates which are measured on each plot, for example, scores of bird damage, weed intensity or disease incidence. The spatial trend is modelled using an ARMA or ARIMA process, as with replicated trials (see Section 5.5). Diagnostics may help with the choice of an appropriate model. Residual errors are assumed to be i.i.d. Normal variables, with mean 0 and variance σ^2.

REML estimates are obtained for σ_g^2, σ^2 and the parameters of the trend model, and are used to derive best linear unbiased predictors (BLUPs) of genotype effects. Spatial analysis for unreplicated trials is available as a procedure within the computer program TwoD (Gilmour, 1992).

6.4.2 Efficiency of spatial analysis

The main advantage of adjusting genotype yields using a spatial model is the coherent framework it provides for incorporating special factors. In the program TwoD, these special factors may include plot covariates, replication of some test genotypes, an irregular pattern of checks and missing plot values. The method automatically derives weights for neighbouring plot values from the estimates of the ratio of genotypic to error variances and spatial correlation structure for the data, so it is not necessary to specify the number of check plots or test genotypes for calculating the environmental index, as required by the other methods. REML estimation also ensures that all information is appropriately weighted in estimating genotype effects.

Cullis, Gleeson and Thomson (1992) conducted a large simulation study to compare the methods of spatial analysis, moving averages and check plots. Table 6.1 compares the weights applied to neighbouring plot yields to estimate genetic effects for the three methods. The frequency of checks was one in five plots. The weights for the BLUPs of genotype yields from spatial analysis depended on the sizes of genetic and trend variance relative to plot variance. The coefficients for the checks are larger than for test genotypes, reflecting the more accurate yield information available on replicated varieties, particularly when the genetic variance is high. In this case, also, the raw yield of the test genotype (g_3) has a large influence on its estimated yield (weights of 0.84, 0.83 in Table 6.1) compared with its weighting when genetic variance is low (0.21, 0.18). The effect of treating test genotypes as random effects is to shrink their yields about the overall mean so that the spread of genotype yields is reduced. Shrinkage is particularly severe when genetic variance is small relative to error variance. In contrast, the weights for moving average and check plot methods are not affected by the size of the genetic or trend variance.

Cullis, Gleeson and Thomson (1992) found that the selection efficiency of spatial analysis was similar to moving averages and superior to using check

Table 6.1. Comparison of the weights for plot yields used to calculate the effect of genotype g_3 in a trial with checks A and B and unreplicated genotypes g_i for different methods of analysis. The weights for the spatial analysis are affected by the size of the genetic component of variance σ_g^2 and trend component σ_t^2 relative to plot error σ^2

Method of analysis	σ_g^2/σ^2	σ_t^2/σ^2	*Weights (×100) for neighbouring genotypes*								
			g_1	g_2	A	g_3	g_4	g_5	g_6	B	...
Spatial	low	high	−2	−3	−6	21	−4	−2	−1	−2	...
	low	low	−2	−2	−3	18	−2	−1	−1	−1	...
	high	high	−4	−4	−44	84	−5	−4	−2	−16	...
	high	low	−3	−3	−23	83	−3	−3	−2	−15	...
Five-point moving average			0	−20	−20	80	−20	−20	0	0	...
Weighted average of checks			0	0	−80	100	0	0	0	−20	...

plots alone. Furthermore, the relative accuracy of the prediction of the genotypic effects for spatial analysis was little affected by check plot frequency, a check plot frequency of one in nine showing the same efficiency as a frequency of one in five.

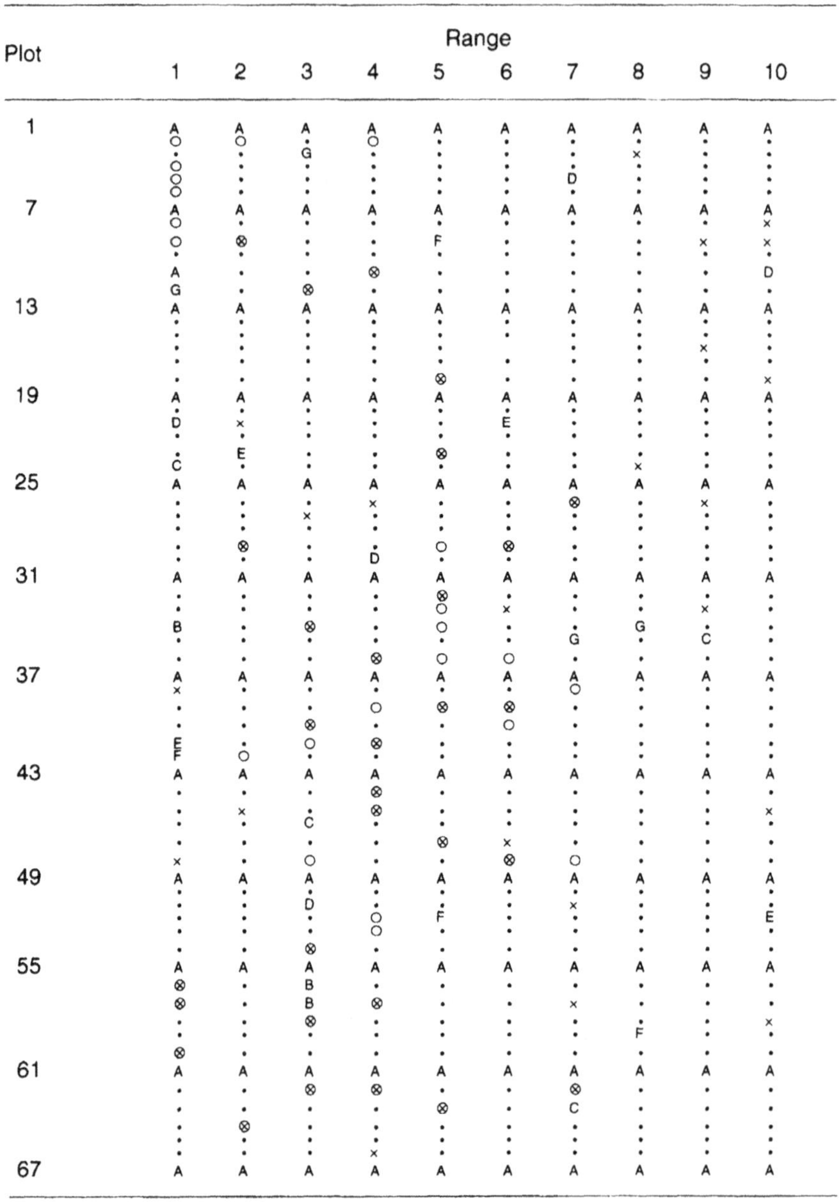

Fig. 6.8. Field layout of the Tullibigeal trial. Replicated check varieties are coded *A* ('Kite') *B* ... *G* and unreplicated genotypes represented by •. The top 10% highest yielding genotypes based on raw data are denoted by o, those predicted by spatial analysis by × (Cullis *et al.*, 1992).

6.4.3 Example of spatial analysis applied to an unreplicated wheat trial

Cullis *et al.* (1989) describe the spatial analysis of an early generation wheat trial conducted at Tullibigeal, New South Wales, Australia, to assess the yield of 525 genotypes. The trial was sown in 10 ranges of 67 plots, each 15 m × 1.8 m (Fig. 6.8) and every sixth plot in each range was sown with the check variety, 'Kite'. Check varieties were also allocated to 25 additional check plots randomly interspersed over the trial. Two yield observations of check plots were missing. The trial was invaded by weeds and a weed score was taken for each plot on a 0 (no weeds) to 10 (total cover) scale.

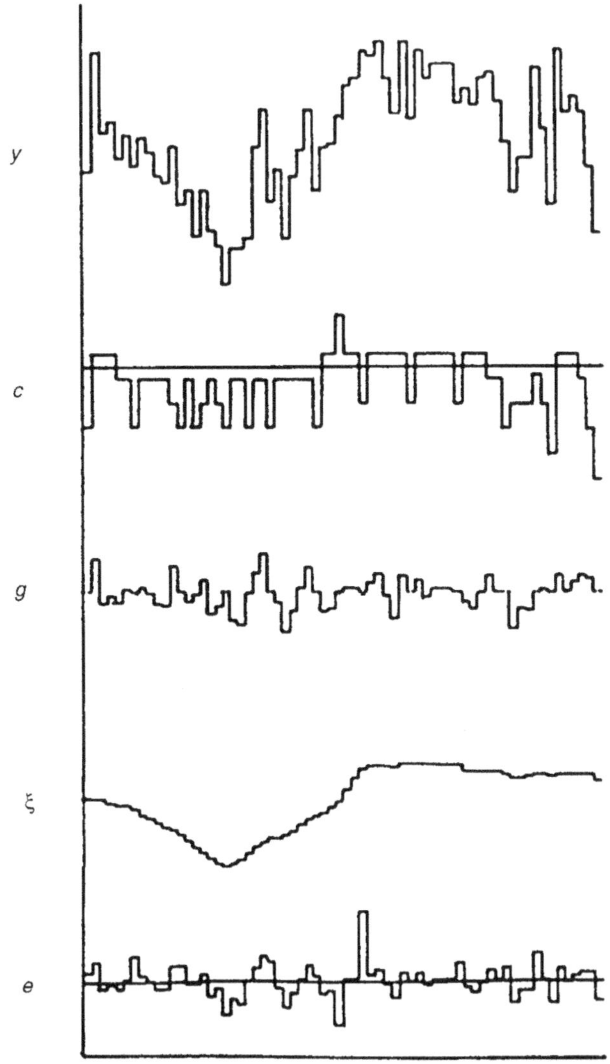

Fig. 6.9. Plot yields y, and components of the yield model, estimated covariate fixed effects c, predicted genotype effects g, trend ξ and local errors e, for range 8 of Tullibigeal trial (Cullis, Gleeson and Thomson, 1992).

In the spatial analysis, an ARIMA (0,1,0) model adequately fitted the trend within ranges and suggested a high level of spatial variation. The weed score covariate was highly significant, with a regression coefficient of -196 ± 22, the negative sign indicating a yield reduction due to weeds.

Figure 6.9 shows the plot yields for range 8, partitioned into the four components of Equation 6.1 by spatial analysis: the fixed effects for weed covariate and checks, predicted effects for test genotypes, spatial trend and plot error. The yield loss from weeds in this range was substantial, but so was the trend after adjusting for the weed score.

The estimated ratio of genotype variance to error variance after spatial analysis was 1.54, large enough to achieve reasonable genetic gain. Figure 6.8 indicates the plot positions of the top 10% of the 525 test genotypes, as predicted by the raw data and the spatial analysis. Thirty of the 52 selections were common to both the raw data and spatial analysis. Using the raw data only, none of the genotypes in ranges 8, 9 or 10 (an area of lower yield) made the top 10% whereas 11 genotypes from these ranges were selected using the predicted values from spatial analysis.

6.5 CONCLUSIONS

Simulation studies by Kempton (1984a) and Cullis, Gleeson and Thomson (1992) indicate that, even without replication, the yields of test genotypes can provide internal information on spatial variation which may improve the efficiency of selection. A requirement of the moving average or more formal spatial analysis approach is that the genotypes are fully randomized across the trial. If genotypes are grouped, say by parentage, then selections may only be made within those groups.

In contrast, check plots provide external information on spatial variation which can be used for adjusting the yields of test genotypes even when these are grouped systematically in the field. However, check plots are costly in terms of resources and may be ineffective if the checks show a different response from the genotypes under test. We would therefore recommend that genotypes are always randomized and check plots only included at low frequency, primarily to aid visual comparison in the field.

7
Interference between plots

R.A. KEMPTON

7.1 INTRODUCTION

Previous chapters have dealt with methods of statistical design and analysis for improving the precision of genotype comparison by controlling sources of environmental variation, particularly those due to soil heterogeneity. Plot interference is another potential source of experimental error which occurs when plot yield is affected not only by the genotype grown in the plot, but also by the particular genotypes in neighbouring plots. Interference is most likely to occur when the plots are small and unbordered, as in the early stages of a plant selection programme. In contrast to plot errors arising from underlying field variation, interference error can lead to a systematic bias in genotype effects that persists across trials and is not reduced by randomization and replication.

Interference frequently arises from competition for resources, such as light, nutrients or moisture. Figure 7.1 shows the effect of interference on the yields of six field bean genotypes with widely differing heights in single-row plots, compared with the centre rows of four-row plots. In single-row plots, yields of taller genotypes are enhanced and shorter genotypes reduced, which leads to a change in genotype ranking compared with the larger plots.

Interference may also be caused by the differential spread of plant pests or pathogens between plots. Thus in a disease screening trial, a highly susceptible genotype may act as a secondary source of infection for other genotypes in the trial so that the effectiveness of more resistant genotypes is underestimated (Parlevliet and van Ommeren, 1984). Figure 7.2 illustrates this for blight scores for four potato genotypes when grown in pure and mixed stands. Although the ordering of the genotypes is the same in both cases, the difference between resistant and susceptible genotypes is much reduced in the mixture. Van der Plank (1963), in his book *Plant Diseases*, called the bias arising from interference 'representational error' as the experimental conditions under which genotypes are assessed are not representative of farming practice.

Statistical Methods for Plant Variety Evaluation. Edited by R.A. Kempton and P.N. Fox. Published in 1997 by Chapman & Hall, London. ISBN 0 412 54750 3.

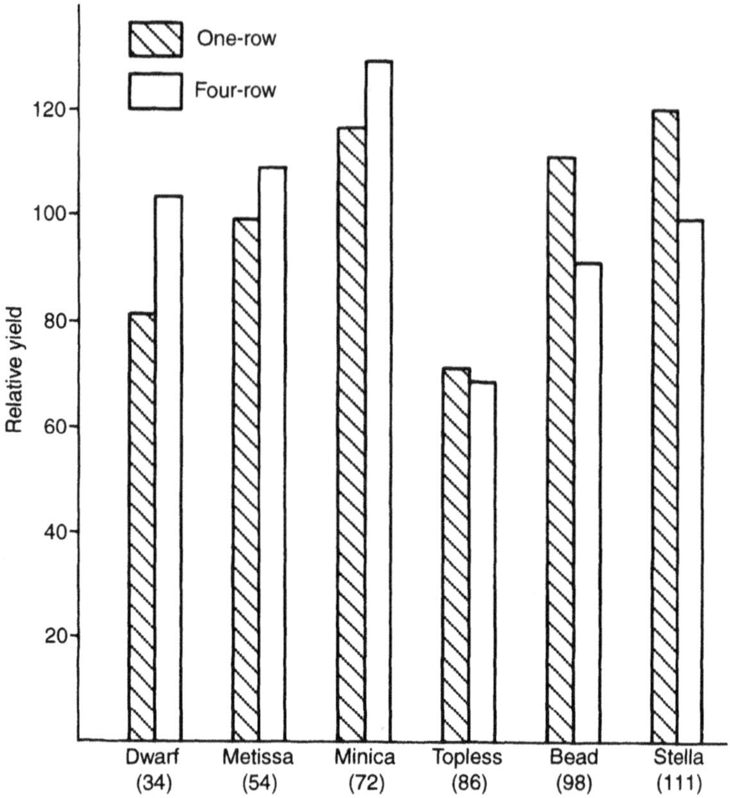

Fig. 7.1. Yields of six field bean genotypes in unguarded single-row plots and the centre rows of four-row plots relative to trial means (= 100) (Kempton and Lockwood, 1984). Figures in brackets are the effective genotype heights in cm.

Spitters (1979) suggested that interference bias is one reason for the poor response to yield selection generally found in early generation trials with small plots. He proposed several methods for reducing the effects of interference: (i) using larger plots and only recording on inner rows; (ii) using wider spacing between plots or separating plots with rows of a standard genotype; (iii) grouping similar genotypes in the field; (iv) applying indirect selection using a character less affected by interference; and (v) applying a mathematical correction for interference. The first three approaches aim to control interference through plot or treatment design. We first consider this aspect in more detail.

7.2 REDUCING INTERFERENCE THROUGH DESIGN

7.2.1 Plot size and borders

Interference can be reduced by increasing plot size or including border plants as guards. In plant breeding trials, however, the scope for increasing plot size

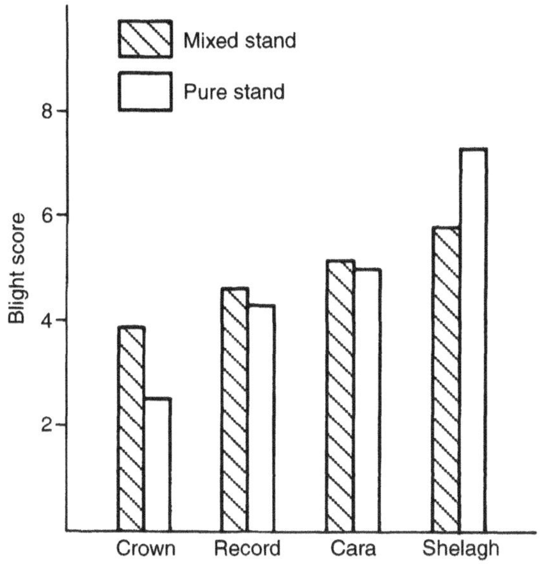

Fig. 7.2. Blight scores (0–9 scale) for four potato genotypes in pure and mixed stands (Kempton, 1992).

is limited by the large numbers of genotypes being assessed and limitations on seed, land and labour. The appropriate choice of plot size and number of border rows will depend on the strength of the interference and how far it extends into the neighbouring plots.

When interference is due to inter-genotypic competition for yield, its effect is often limited to the outer row of adjacent plots, though smaller but significant effects on the second row have also been observed by Gomez (1972) in rice, Rich (1973) in wheat and Kempton and Lockwood (1984) in field beans. Indeed, Wiancko, Arny and Salmon (1921) in their 'Report of the Committee on Standardisation of Field Experiments' recommended that, for small grain crops, two border rows be used either side of each plot. In practice, however, the effect of inter-plot competition will be largely removed by using a single border row.

Several studies have compared the effect of plot width and use of border rows on inter-plot competition. Kramer, van Ooijen and Spitters (1982) compared the yields of 16 spring wheat genotypes in one-, three- and six-row plots with the yields in large plots (Table 7.1). For whole plots without borders, the coefficient of variation decreased and the correlation with genotype yields in the large field plots increased with increasing plot width, reflecting the greater area sampled and the reduction in competition from neighbours. Bordered plots obtained from harvesting the centre rows of the three- or six-row plot trial gave the same correlation as harvesting the whole plot. Here the smaller bias in bordered plots is balanced by a larger random error due to the smaller area sampled, particularly for three-row plots.

Where it is impracticable to harvest less than the full plot width, a more

Table 7.1. Coefficient of variation of the yields of wheat genotypes in small plots of different types, and correlation with yields in standard field trial (from Kramer, van Ooijen and Spitters 1982)

Plot type	*Coefficient of variation*	*Phenotypic correlation*
Single-row		
standard spacing	0.28	0.33
double spacing	0.15	0.56
Three-row		
complete plot	0.09	0.74
central row	0.21	0.74
Six-row		
complete plot	0.08	0.89
central four rows	0.08	0.88

expensive alternative is to use a full plot as a guard, with the right and left halves sown with the genotypes of the right and left hand test plot. When seed of the test genotypes is limited, border plots are sometimes sown with a standard genotype of intermediate competitive ability (Spitters, 1979), or wider alleys are used (Kramer, van Ooijen and Spitters, 1982; Fasoulas, 1990) to separate test plots. However, this will be only partially effective and may indeed introduce other sources of error (see Chapter 2).

Where disease is a factor, interference may often extend beyond the adjacent plots. The spread of disease pathogens can be studied through the distribution of infected plants about a point source of disease (Waggoner, 1952; Jenkyn and Bainbridge, 1974). For aerially dispersed pathogens, wind speed and direction over the period of disease development is a major factor. Paysour and Fry (1983) derived a model for pathogen spread to study how plot size and distance between the plots affect interference. Experiments have also been carried out with resistant and susceptible genotypes for a number of crops and diseases. Examples include leaf rust on barley (Parlevliet and van Ommeren, 1984), wheat (Bowen, Teng and Roelf, 1984) and maize (Randle, Davis and Groth, 1986); and powdery mildew on wheat (Lipps and Madden, 1992). In general, smaller plots without guard rows showed larger interference and a smaller range of resistance scores among genotypes compared with larger, more widely separated plots.

7.2.2 Grouping similar genotypes

Interference may also be reduced by grouping similar genotypes in the field. If the grouping is successful, the interference experienced by a genotype from similar genotypes in neighbouring plots will be close to that for the genotype in a pure stand. In the early stage of selection, there may be little prior information on the genotypes and the best strategy then is to group genotypes from the same cross. Alternatively, Kaufmann and McFadden (1960) recommended grouping by seed size as this can affect juvenile vigour and, consequently, competitive ability. For advanced germplasm, where more information is available, grouping may be based on other plant characters likely to affect interference: for example, height, maturity or reactions to diseases.

Each group of genotypes may be grown in separate trials but this will restrict statistical comparisons between different groups. An alternative is to use an incomplete block design and arrange the genotypes so that those from different interference groups never occur together in the same block. Genotypes from different groups can then only occur as neighbours at block boundaries, and, if necessary, blocks can be bordered to avoid this. One disadvantage with this design is that comparisons between genotypes in different interference groups are based wholly on between-block information and will therefore be estimated with poor precision when the trial site is very heterogeneous (see Section 4.6). Consider, for example, the design in Table 7.2(a) with three replicates of 20 genotypes allocated to three interference groups of sizes 4, 12 and 4. A block size of 4 is used. Genotypes from groups 1 and 3 are allocated to blocks I and V respectively, while genotypes from group 2 are arranged among blocks II, III and IV using an alpha design. Let the variance component for plots within blocks be σ^2, and that between blocks be σ_b^2. Then the variance of differences in means for genotypes within group 1 or within group 3 (i.e. from the same block) is $\sigma^2/3$ while the corresponding variance for genotypes from different groups (i.e. from different blocks) is $(\sigma^2 + 4\sigma_b{}^2)/3$. Thus if $\sigma_b^2 = \sigma^2$, a typical situation for UK cereal trials, the variances of genotype differences range from $0.33\sigma^2$ to $1.67\sigma^2$ with this design. In contrast, if the grouping of genotypes is ignored and a standard alpha design used with five blocks of four plots per replicate, then, using only within-block information, individual variances have a range $0.42\sigma^2$ to $0.47\sigma^2$, and the range will be even smaller if between-block information is used. Hence, for the grouped design, the potential benefit of reduced bias in genotype comparisons is achieved at the potential cost of increased variance compared with a standard alpha design. In practice, the choice of design will depend on the relative importance of interference and spatial heterogeneity for the trial.

When interference groups can be ordered, for example by height, David and Kempton (1996) propose compromise designs which eliminate the more extreme forms of interference but still maintain similar precision to alpha designs. They characterized a group divisible design for c groups by its set of group concurrences $\{\lambda_{uv}, u, v = 1 \ldots c\}$, where λ_{uv} is the average number of times pairs of genotypes from groups u and v occur in the same block. They then investigated designs with high overall efficiency for which genotypes from extreme interference groups appeared in the same block with only small probability: i.e. λ_{uv} is small when $|u-v|$ is large. Table 7.2(b) gives a design for 20 genotypes as before, in which there are two extreme interference groups of four genotypes and a larger intermediate group of 12. Genotypes are allocated to blocks so that those from the extreme groups

Table 7.2. Incomplete block designs with blocks of four plots for 20 genotypes allocated to three ordered interference groups with 4, 12 and 4 genotypes

(a) Genotypes from different groups occur in different blocks					
Replicate 1					
block	I	II	III	IV	V
genotype	(1 2 3 4)	(5 6 7 8)	(9 10 11 12)	(13 14 15 16)	(17 18 19 20)
group	1 1 1 1	2 2 2 2	2 2 2 2	2 2 2 2	3 3 3 3
Replicate 2					
block	VI	VII	VIII	IX	X
genotype	(1 2 3 4)	(5 10 15 8)	(9 14 7 12)	(13 6 11 16)	(17 18 19 20)
group	1 1 1 1	2 2 2 2	2 2 2 2	2 2 2 2	3 3 3 3
Replicate 3					
block	XI	XII	XIII	XIV	XV
genotype	(1 2 3 4)	(5 14 11 12)	(9 6 15 16)	(13 10 7 8)	(17 18 19 20)
group	1 1 1 1	2 2 2 2	2 2 2 2	2 2 2 2	3 3 3 3
(b) Genotypes from extreme groups 1 and 3 occur in different blocks					
Replicate 1					
block	I	II	III	IV	V
genotype	(1 2 5 6)	(3 4 7 8)	(9 10 11 12)	(13 14 17 18)	(15 16 19 20)
group	1 1 2 2	1 1 2 2	2 2 2 2	2 2 3 3	2 2 3 3
Replicate 2					
block	VI	VII	VIII	IX	X
genotype	(1 3 14 9)	(2 4 10 16)	(5 8 13 15)	(7 12 17 19)	(6 11 18 20)
group	1 1 2 2	1 1 2 2	2 2 2 2	2 2 3 3	2 2 3 3
Replicate 3					
block	XI	XII	XIII	XIV	XV
genotype	(1 4 13 12)	(2 3 11 15)	(6 7 16 14)	(5 9 17 20)	(8 10 18 19)
group	1 1 2 2	1 1 2 2	2 2 2 2	2 2 3 3	2 2 3 3

never occur together in the same block. The group concurrence matrix for the design is

$$\Lambda = \begin{bmatrix} \lambda_{11} & & \\ \lambda_{12} & \lambda_{22} & \\ \lambda_{13} & \lambda_{23} & \lambda_{33} \end{bmatrix} = \begin{bmatrix} 1.0 & & \\ 0.5 & 0.45 & \\ 0 & 0.5 & 1.0 \end{bmatrix}$$

In contrast, for a standard incomplete block design ignoring groups, the average number of times any pair of genotypes occurs in the same block is 0.47. The new design eliminates the concurrence of extreme genotypes (from groups 1 and 3), while increasing concurrences within these groups; concurrences involving the intermediate group 2 are little affected. The design will thus be effective at reducing the bias from interference.

The design is also effective in controlling heterogeneity; pairs of genotypes do not occur together more than once in the same block, and genotypes which never occur together are well compared through indirect comparisons (see Chapter 4). For groups 1 and 3, each genotype appears once in the same block with every other genotype from the same group; for group 2, each genotype appears once in a mixed block with genotypes from group 1, once in a block with genotypes from group 3 and once in a block with only genotypes from its own group. Hence, genotypes from groups 1 and 3 can be compared indirectly through comparisons with genotypes from group 2. David and Kempton (1996) show that this design has an efficiency of 0.742, only slightly less than the maximum efficiency of 0.745 achieved with an alpha design. Ignoring between-block information, the variance of genotype differences range from $0.42\sigma^2$ for comparisons within groups 1 or 3, to $0.49\sigma^2$ for comparisons between these extreme groups.

For the above designs, the restriction on randomization only affects the allocation of genotypes to interference groups. The three stages of randomization are as follows (cf. Section 4.5.5): (i) random allocation of genotypes to codes within groups; (ii) random allocation of blocks within replicates; (iii) random allocation of genotypes to plots within each block.

Where interference is restricted to adjacent plots, however, we might consider restricting the randomization of genotypes within and across blocks to reduce interference effects. An extreme example is a systematic design in which genotypes are ordered, for example by height, in each replicate; Cleaver, Greenwood and Wood (1970) describe the use of such designs for fertilizer experiments. This reduces interference and the need for guard plots, but a lack of randomization may lead to biased estimates of genotype differences and their standard errors due to fertility trends. David, Kempton and Nevison (1996) propose modified alpha designs in which the randomization is restricted so that pairs of genotypes with large interference never appear in adjacent plots. This is illustrated in Table 7.3 which gives the field layout of a design for comparing 20 genotypes from three ordered interference groups, as above. Replicates consist of separate lines of plots. The restricted randomization is based on cyclic permutations of genotypes within blocks and ensures that all genotypes have an equal probability of occurring on each plot, while genotypes from groups 1 and 3 never appear as neighbours. The designs can be generated using the computer program ALPHA+ (Williams and Talbot, 1993).

Table 7.3. Field layout of modified alpha design with three replicates and blocks of four plots, for comparing 20 genotypes associated with three ordered interference groups defined as in Table 7.2. | indicates block boundary

Replicate 1																				
	11	*20*	*14*	*3* │	*7*	*16*	*10*	*4* │	*5*	*6*	*9*	*1* │	*12*	*18*	*19*	*15* │	*8*	*17*	*13*	*2*
	2	3	2	1	2	2	2	1	2	2	2	1	2	3	3	2	2	3	2	1
Replicate 2																				
	20	*12*	*2*	*9* │	*14*	*4*	*5*	*17* │	*19*	*8*	*1*	*10* │	*15*	*3*	*7*	*6* │	*13*	*18*	*16*	*11*
	3	2	1	2	2	1	2	3	3	2	1	2	2	1	2	2	2	3	2	2
Replicate 3																				
	19	*7*	*2*	*14* │	*15*	*1*	*11*	*17* │	*8*	*16*	*3*	*9* │	*4*	*12*	*6*	*13* │	*20*	*18*	*10*	*5*
	3	2	1	2	2	1	2	3	2	2	1	2	1	2	2	2	3	3	2	2

The success of group divisible designs depends on the effectiveness of the allocation of genotypes to interference groups. If the grouping is unreliable (Christidis, 1939), using these designs could lead to a larger interference than with a standard design which ignores genotype groupings.

7.3 MODELLING INTERFERENCE

Where it is not practicable to remove interference through design, it may be possible to take account of its effects in the statistical analysis by including additional terms in the response model. Such models have been used to analyse competition diallels (Williams, 1962; McGilchrist, 1965; Trenbath, 1978).

The basic diallel model is introduced here for an experiment with four genotypes of field bean (Kempton and Lockwood, 1984). Each plot consisted of three, four-row subplots, made up of a central harvested subplot of one genotype and two adjacent subplots of an associate genotype. There were four replicates of each of the 16 genotype combinations. The yield of the centre plot of genotype *A* with genotype *B* as neighbour is modelled as

$$\begin{array}{ccccccccccc} \text{yield } A & = & \text{replicate} & + & \text{direct} & + & \text{interference} & + & \text{interaction} & + & \text{plot} \\ \text{with } B & & \text{mean} & & \text{effect } A & & \text{effect } B & & A \text{ with } B & & \text{error} \end{array} \quad (7.1)$$

where the direct effect of a genotype is its effect on plot yield and the interference effect is its effect on neighbouring plots, averaged over all genotype neighbours.

Table 7.4 gives the mean yield of each genotype with each of the four genotypes as neighbours. The margins of the table are estimates of the direct and interference effects. The residuals in the body of the table are estimates of the interaction effects: for example, the interaction effect for 'Dwarf' when grown with 'Topless' is calculated as 1847 − 1633 − 31 − 90 = 93. Overall, the genotypes 'Dwarf' and 'Topless' show positive interference effects, while 'Maris Bead' and 'Stella Spring' have negative effects. Despite the sizeable residual effects in Table 7.4, Kempton and Lockwood (1984) found that genotype by neighbour interactions were not significant when compared against plot error.

Table 7.4. Components of yield for a competition diallel experiment with four genotypes of field bean. The mean plot yields in g/m^2 for four replicates (bold figures) are partitioned into a grand mean, a direct effect, an interference or neighbour effect and a residual

Genotype	*Neighbour*				*Direct effect*
	Dwarf	*Topless*	*Maris Bead*	*Stella Spring*	
Dwarf	**1829**/42	**1847**/93	**1506**/−56	**1470**/−83	+31
Topless	**1262**/−24	**1252**/−1	**991**/−70	**1145**/+93	−470
Maris Bead	**1687**/−120	**1758**/−16	**1674**/+92	**1614**/+41	+51
Stella Spring	**2245**/101	**2033**/−78	**1950**/+31	**1858**/−52	+388
Interference effect	+123	+90	−102	−111	
		Grand mean 1633			

Equation 7.1 can be extended to experiments where each plot has a number of different genotypes as neighbours. Suppose the direct yield effect of genotype i is g_i, and the interference effect of a neighbouring genotype k on genotype i is ϕ_{ki}. Then, the yield of the ith genotype in the jth plot may be written as (Kempton, 1985)

$$y_{ij} = \mu + b_j + g_i + \sum_k \phi_{ki} w_{jk} + e_{ij} \tag{7.2}$$

where μ is the trial mean, b_j is the block effect corresponding to the jth plot and w_{jk} is the number of times genotype k is in a neighbouring plot to plot j. When the interference is restricted to adjacent plots in one dimension, $w_{jk} = 1$ if genotype k is in plot $j - 1$ or $j + 1$, otherwise $w_{jk} = 0$. However, for the longer range interference found in disease trials, it may be more appropriate to define all of the plots in the same block as neighbours. The model can be further extended by giving different weights to neighbours at different distances from each plot, or by defining separate interference effects for left and right hand neighbours to take into account, for example, the direction of the sun or the prevailing wind. Equation 7.2 allows for non-additivity between direct effects and interference effects. If additivity is assumed, then ϕ_{ki} can be replaced by the average interference effect of genotype k, ϕ_k.

7.3.1 Designs for estimating interference effects

The number of plots required for a complete diallel design makes them impracticable even for moderate numbers of genotypes. When interference is restricted to adjacent plots, and effects are additive, they may be estimated using a more concise neighbour design. In a balanced neighbour design, each genotype occurs as a neighbour to every other genotype an equal number of times. Table 7.5 gives examples of designs for $g = 7$ genotypes: balanced designs of type (a) generally have $\frac{1}{2}(g - 1)$ replicates and are only available for odd numbers of genotypes; design (b) with $(g - 1)$ replicates is separately balanced for both left and right hand neighbours; design (c) is balanced in two dimensions but requires a large number of border plots. The pairwise neighbour balance means that direct and neighbour effects are orthogonal, and leads to efficient estimates of pure stand effects. Balanced designs with up to 15 genotypes and seven replicates have been used by Austin, Ford and Morgan (1989), but large numbers of genotypes will usually only permit a partially balanced neighbour design. Extensions of neighbour designs to include equal numbers of genotype triplets for investigating the interactions between direct and interference effects (Dyke and Shelley, 1976; Lin, Poushinsky and Voldeng, 1985) require too many plots for general use.

Table 7.5. Balanced neighbour designs for seven genotypes with border plots

(a) In one dimension with two neighbours (Williams, 1952)
G (A B C D E F G) (B D F A C E G) (C F B E A D G) A

(b) In one dimension with separate balance for left and right hand neighbours (Williams, 1952)
C (A B C D E F G) (C E G B D F A) (C F B E A D G) (F ...
... G) (F C G D A E B) (F E D C B A G) (A F D B G E C) A

(c) In two dimensions with four neighbours (Freeman, 1979)

F	*A*	*G*	*C*	*E*	*B*	*D*	*F*	*E*
G	*B*	*D*	*F*	*A*	*G*	*C*	*E*	*D*
D	*A*	*C*	*E*	*G*	*F*	*B*	*D*	*A*
E	*D*	*F*	*A*	*C*	*B*	*E*	*G*	*B*
A	*C*	*E*	*B*	*F*	*G*	*A*	*D*	*C*

Where all of the plots within the same block are defined as neighbours, interference effects may be estimated using a balanced incomplete block design (Chapter 4). Once again, when the number of genotypes is large, achieving complete balance is impracticable and a partially balanced design should be used.

7.3.2 Relating interference effects to plant characters

The interference effect of a genotype on yields of neighbouring plots is related to its phenotype. Many studies have found a close relationship between interference and plant height: in cereals (Jensen and Federer, 1964; Fischer, 1979; Kempton *et al.*, 1986); in cassava (Kawano *et al.*, 1978); in field beans (Kempton and Lockwood, 1984); in kale (Bradshaw, 1986); and in rice (Gomez, 1972). For example, Fig. 7.3 plots the estimates of interference effects for old and new genotypes of winter wheat whose heights ranged from 71 cm to 155 cm (Austin, Ford and Morgan, 1989). For UK variety trials, Talbot *et al.* (1995) showed a significant relationship between the interference

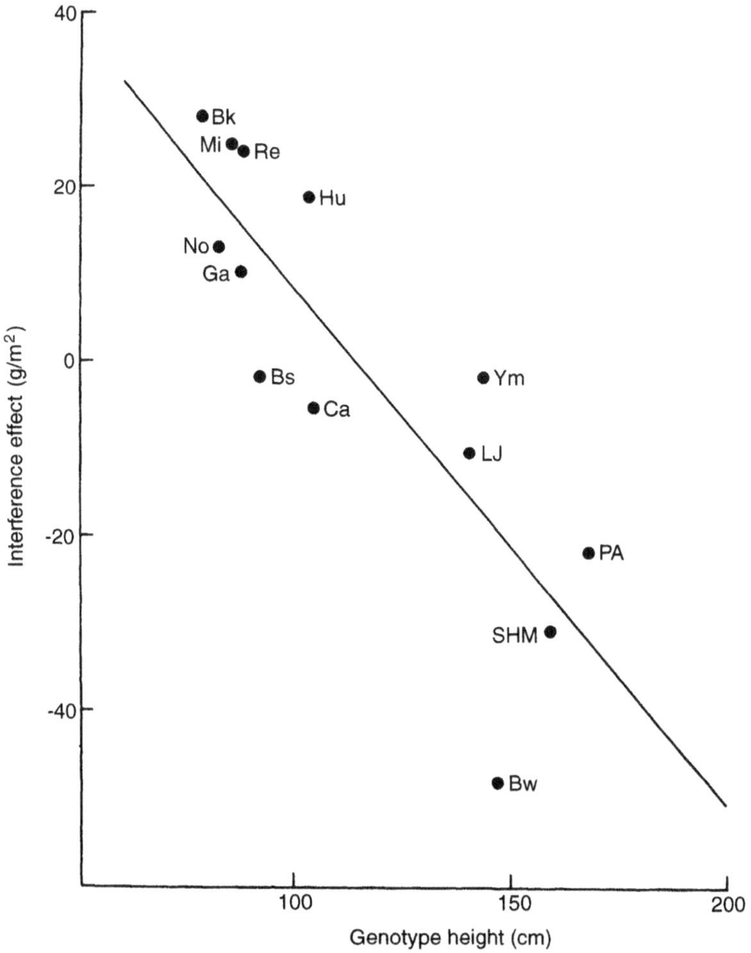

Fig. 7.3. Relationship between the interference effect for yield and plant height for 13 winter wheat genotypes (for genotype names, see Kempton, 1985).

effects and height in winter barley, winter wheat, winter oilseed rape and forage maize. Interference effects were also related to % lodging in both spring and winter wheat and oilseed rape, to % mildew in winter wheat and oilseed rape, to % brown rust in winter barley, and to days to maturity in winter wheat. In root crops, several authors have observed interference associated with root yield: in sugar beet (Kempton, 1982); in potatoes (Kempton, 1985; Connolly *et al.*, 1993); and in swedes (Bradshaw, 1989). Talbot *et al.* (1995) also found a significant relationship between the interference effects and top size in sugar beet. Such relationships can be used to simplify interference models.

7.4 ADJUSTING FOR INTERFERENCE

Equation 7.2 provides estimates of the genotype effects adjusted for bias from interference. The model is reparameterized so that the interference experienced by each genotype is standardized to that expected with itself as neighbour:

$$y_{ij} = \mu + b_j + g_i^* + \sum \phi_{ki}^* w_{jk} + e_{ij} \qquad (7.3)$$

where g_i^* is the effect of genotype i in the pure stand and ϕ_{ki}^* is the average difference in response with genotype k rather than genotype i as neighbour. Assuming additivity between direct and neighbour effects, $\phi_{ki}^* = \phi_k - \phi_i$. Then, if each plot has m neighbours which are given equal weights, the pure stand effect for genotype i is $g_i^* = g_i + m\,\phi_i$. Equation 7.3 can be fitted to data by least squares to give estimates of the genotype pure stand effects and their standard errors. Where incomplete blocks are used, inter-block information on the genotype and interference effects is often important (Brown and Kempton, 1994).

When the number of replicates is small, however, the average interference effects ϕ_k are poorly estimated and consequently estimates of the genotype pure stand effects g^* have large standard errors. One way of reducing the number of interference parameters is to allocate genotypes to interference groups, as in Section 7.2.2, and to assume that all of the genotypes in a group have the same interference effect. Further parameter reduction is possible when the interference is associated with a particular plant character (Section 7.3.2). In particular, if the interference effect is linearly related to the mean character score X for a genotype (e.g. plant height, as in Fig. 7.3), then $\phi_{ki}^* = \beta(X_k - X_i)$, and the interference can be modelled by a single linear covariate with a regression coefficient β. As a variant of the above, the covariate X may be taken from the observed character score in the neighbouring plot, rather than the genotype mean. The response of the jth plot can then be written as

$$y_{ij} = \mu + b_j + g_i^* + m\beta(X_{N(j)} - X_j) + e_{ij} \qquad (7.4)$$

where $X_{N(j)}$ is the mean score of the m neighbours of plot j.

A special case is when the interference effect is proportional to the genotype response, which is sometimes appropriate in disease trials (Kempton, 1992) or yield trials with root crops (Kempton, 1982). Then $\phi_i = \beta g_i^*$ and thus $g_i^* = g_i/(1 - m\beta)$. Now, for a balanced neighbour design, g_i is estimated by the genotype mean. Hence, in this case, the adjustment to the pure stand is achieved by a simple scaling and the ordering of the genotypes is not affected by interference (see Fig. 7.2). This led Parlevliet and van Ommeren (1984) to suggest calibrating the scores from a disease nursery using scores for standard genotypes grown in large plots. Figure 7.4 shows the adjustment of disease scores for ten hypothetical genotypes. The genotypes are grown in small plots with four standard genotypes which span the full range of disease resistance. The resistance for the standards is also assessed in large, isolated plots. A linear regression of small plot score on large plot score is fitted for the four standards, and then used to adjust the scores for the other genotypes. In some cases a non-linear calibration curve may be more appropriate.

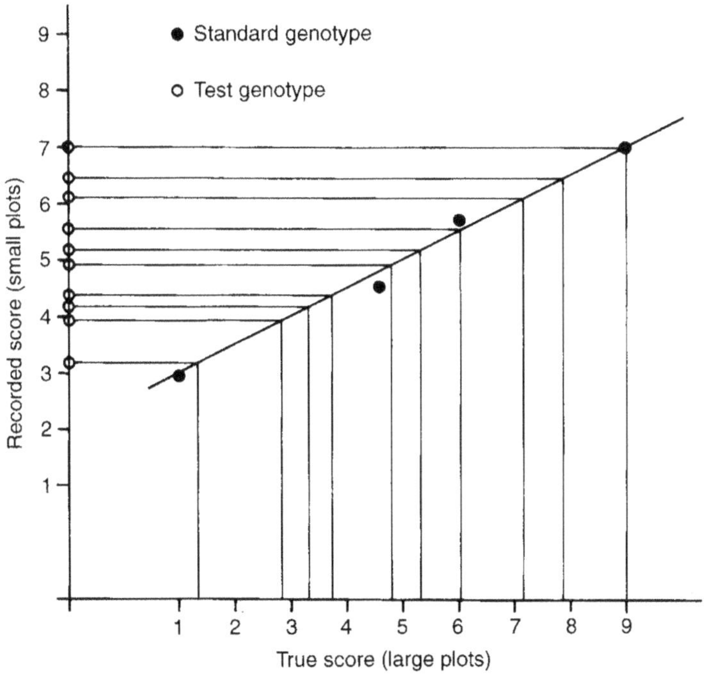

Fig. 7.4. Adjustment of disease score for ten test genotypes in small plots using a calibration regression equation derived for four standard genotypes.

7.4.1 Fitting the interference model to the field bean trial

We illustrate the method of model selection and adjustment for interference for a trial of six field bean genotypes grown in single-row plots (Kempton and Lockwood, 1984). A balanced neighbour design was used in which each genotype also had itself as a neighbour at the boundary of a replicate [compare with the design of Table 7.5(b)]. The design was based on four repeats of the following linear arrangement of 36 plots, with ends bordered:

B (*B D A E C F*) (*F E D C B A*) (*A D B F C E*) (*E B C A F D*) (*D F B E A C*) (*C D E F A B*) *B*

The full design with plot yields and heights is given by Besag and Kempton (1986).

Plot yields were analysed, excluding border plots, and the interference was assumed to be restricted to adjacent plots. The fit of Equation 7.3 with additive interference effects was first compared with the null model [Table 7.6(i) and (ii)] and interference effects were found to be significant ($F_{5,110} = 2.73$, $P = 0.023$). To test for the additivity of interference effects and direct effects, the full Equation 7.3 was fitted, although the design permitted only 18 of the 35 interference terms to be estimated [Table 7.6(iii)]. Comparison with the fit of the additive model showed no strong evidence of interaction ($F_{13,97} = 1.76$, $P = 0.061$) so the additive model was adopted. Genotype effects in pure stands are then given by $g^* = g + 2\phi$, where, because the design is balanced for neighbours, the estimates of direct effects g are derived directly from the unadjusted genotype means (Table 7.7).

Table 7.6. Fit of interference models to field bean yields measured by the residual sum of squares (RSS) and residual mean square (RMS)

Model	*df*	*RSS*	*RMS*
(i) No interference	115	633063	5505
(ii) Additive interference effects	110	563099	5119
(iii) Interaction of interference and direct effects	97	455561	4697
(iv) Interference effects proportional to height difference of neighbours	113	552626	4890

Table 7.7. Estimates of the yields for six genotypes of field beans in single-row plots using different interference models, and comparison with the centre rows of an adjacent four-row trial

Genotype (height cm)	*Single-row trial (mean yield, 391 g/row)*			*Four-row trial (mean, 503 g/row)*
	Ignoring interference	*Adjusted for interference*		*Observed centre-row effects*
		Equation 7.3 (ϕ)	*Equation 7.4*	
Stella Spring (111)	+79	+57 (−11)	+45	−4
Minica (72)	+65	+93 (+14)	+74	+148
Maris Bead (98)	+43	−1 (−22)	+18	−47
Metissa (54)	−3	+21 (+12)	+24	+45
Dwarf (34)	−72	−23 (+25)	−20	+17
Topless (72)	−112	−148 (−18)	−141	−159
SED	21	38–47	20–35	16

For Equation 7.3, interference effects ϕ are assumed to be additive and their estimates appear in brackets; for Equation 7.4, interference effects are assumed proportional to crop height

The interference effects are approximately related to the heights of the genotypes, so a covariance analysis was applied using Equation 7.4 with the covariate X_j equal to the crop height in plot j. An additional adjustment was also made for the aberrant behaviour of the genotype 'Topless' which has a large negative interference effect despite its average height. This can be explained by the genotype's determinate growth habit (Kempton and Lockwood, 1984) which reduces its final height, though for the majority of the growing period it has similar height to the taller genotypes, 'Stella Spring' and 'Maris Bead'. A separate interference effect was therefore included in Equation 7.4 to adjust the height of 'Topless'. The estimated covariate regression coefficient on height difference was $\hat{\beta} = 0.69 \pm 0.195$ g/cm and the effective height of 'Topless' as a competitor was estimated to be 14 cm greater than its final observed height. This model fits the data better than the model with separate interference effects, despite having fewer parameters (Table 7.6). This indicates that greater precision is achieved using actual plot heights to model local interference. The estimates of genotype effects in pure stands for both models are similar (Table 7.7), but the SEDs for Equation 7.4 are considerably smaller.

The need to adjust yields from the single-row plots is supported by results from the centre rows of an adjacent four-row trial with the same genotypes (Table 7.7). Interference between plots affects the ordering of genotypes, with taller genotypes showing considerable gains in yield compared with their shorter neighbours. This distortion is only partly corrected by the analysis of covariance.

7.5 SUMMARY

There is extensive evidence that interference among genotypes in neighbouring plots can bias comparisons in plant breeding trials. As illustrated in Fig. 7.2, interference sometimes affects only the scale of responses, while, at other times (Fig. 7.1), it changes the ordering of genotypes. The size of interference effects will depend on the plot size and the variation of the plant material. Interference effects for yield have been related to a number of plant characters: for root crops, an important factor is the root yields of neighbouring plots; while for cereals, interference is commonly associated with differences in plant height. For example, results from cereal experiments with 1.5 m wide plots (Kempton *et al.*, 1986) indicate that the plot yield will be increased by an average of 0.2% for every 1 cm a genotype is taller than its neighbours; shorter genotypes will show a corresponding decrease in yield.

Interference is most effectively dealt with by the increasing plot size (more specifically, plot width) or by omitting border plots from yield assessment. In many breeding programmes, however, this is impracticable and alternative methods must be considered. Other design approaches based on grouping similar genotypes require prior knowledge about genotype traits which may not be available in early generations. In this case, the best approach will usually be to adjust for interference using an appropriate plant covariate derived from neighbouring plots. Interference models can also be used to test for the presence and size of interference effects. These tests should be repeated whenever there is a significant change in the genetic material or field plot technique.

8

Multi-environment testing and genotype × environment interaction

P.N. FOX, J. CROSSA AND I. ROMAGOSA

8.1 WHAT IS GENOTYPE × ENVIRONMENT INTERACTION?

Genotype by environment interaction (GE) is differential genotypic expression across environments. This chapter summarizes the implications for breeding of the reduction in association between phenotypic and genotypic values caused by GE. Statistical techniques for quantifying genotypic adaptation and for characterizing environments are presented, focusing mostly upon analyses of two-way tables of genotype by environment data. Genotype estimates for different environments in the same GE table may be derived using different methods of trial design and analysis.

Means across environments are adequate indicators of genotypic performance in trials with non-significant GE. However, when GE is significant, these means often mask subsets of environments where genotypes differ markedly in relative performance. Selections from one environment may often perform poorly in another. Plant breeders therefore consider genotypic adaptation across environments. Statistical analyses of multi-environment trials should detect GE, quantify it and focus upon interpretable GE variation to help determine a breeding strategy for adapted genotypes.

GE is the major element in determining many key aspects of a breeding programme including (i) whether to aim for wide or specific adaptation; (ii) choice of locations for selection; (iii) whether selection in early generations is conducted in stressed or stress-free environments; and (iv) the trade-off between multi-environment testing of large numbers of genotypes and subjecting fewer lines to intensive trait-based selection. Breeders generally consider genotypic adaptation across environments through an informal synthesis of observations or through more rigorous statistical measures of the stability of individual genotypes.

Becker (1981) distinguished two types of genotypic stability. Biological or

Statistical Methods for Plant Variety Evaluation. Edited by R.A. Kempton and P.N. Fox. Published in 1997 by Chapman & Hall, London. ISBN 0 412 54750 3.

homeostatic stability, in which a genotype maintains a constant yield across environments, is assessed by genotypic variance across environments. Consideration of a genotype's ability to respond to improved conditions underlies the concept of agronomic stability, by which a genotype is considered stable if it yields well relative to the productive potential of the test environments. Implicitly, the idea of agronomic stability is related to ranking. A stable genotype is one that is consistently well ranked. If agronomic stability is demonstrated for a wide range of environments, a genotype is deemed to have general or wide adaptation. If agronomic stability is confined to a limited range, a genotype has specific or narrow adaptation.

We use stability and adaptation to refer to both spatial and temporal dimensions, as most studies assume stability measured spatially and temporally to be expressions of the same phenomenon. However, the assumption that spatial and temporal buffering rely on the same genetic mechanisms has rarely been tested. Many stability parameters are correlated with genotype mean yields and so contribute little to plant variety improvement. This correlation may frustrate the development of methodology to quantify adaptation while, on the other hand, it may signify that selection for mean yield will achieve some level of favourable adaptation.

This chapter describes the current methodology for improving adaptation above what is achievable by selection for mean yield within a breeder's individual domains for release and recommendation. The next section introduces the concept of forming such domains. We have not presented an historical review of the subject nor fully discussed the labyrinthine literature on GE. Our approach emphasizes options for today's breeding programmes without providing full details of the statistical techniques.

8.2 IMPLICATIONS FOR BREEDING PROGRAMMES

In plant breeding, the most important GE is of the 'crossover type' (Fig. 8.1), which implies changes in the rankings of genotypes across environments (Baker, 1988). With non-crossover interactions, genotypes with superior means can be recommended for all environments.

For trials in which the locations and genotypes are included across the years, the GE term from the analysis of variance can be partitioned (see Chapter 10) into components due to genotypes × locations (G · L), genotypes × years (G · Y) and genotypes × locations × years (G · L · Y). If G · L is the dominant portion of GE, then specific adaptation is exploitable by subdividing the production areas into homogeneous (but not necessarily contiguous) regions that minimize GE within regions and form domains for release and recommendation of genotypes. Subdivision is conditional upon the genotypes employed, and, as the germplasm in a programme evolves, the subdivisions may change.

Where G · Y and G · L · Y terms dominate, no simplification involving spatial subdivision of breeding regions is available. However, it is important that trials cover a representative range of conditions to determine the genotypic responses.

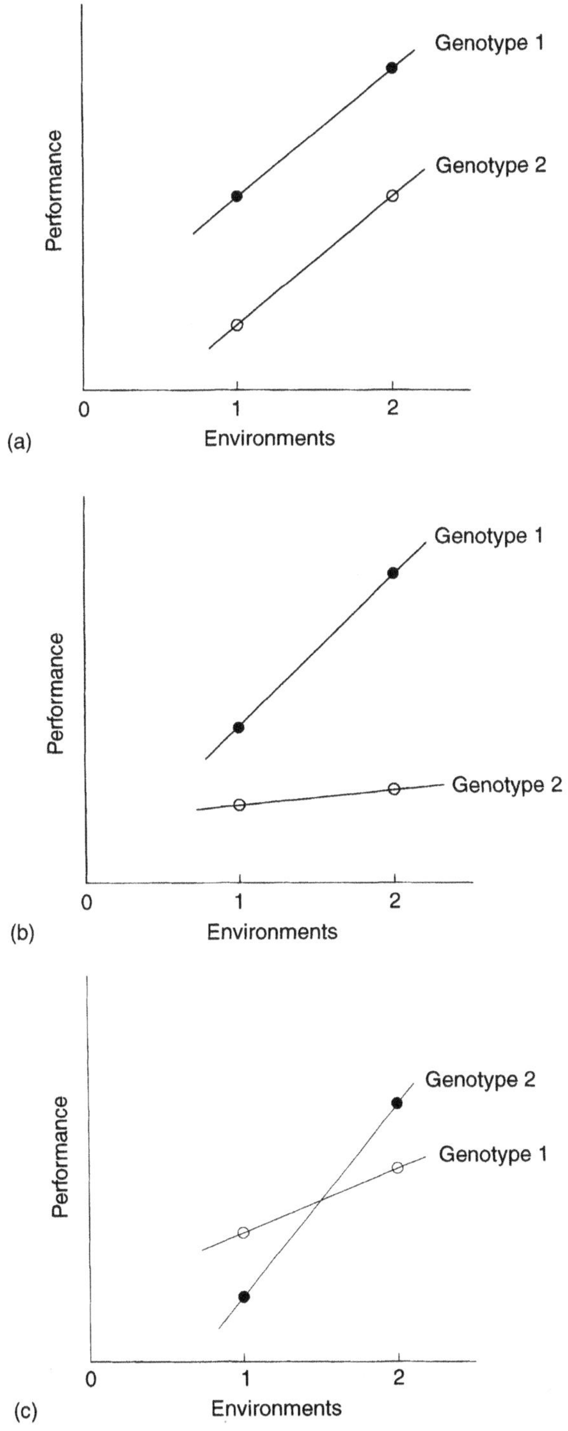

Fig. 8.1. The performance of two hypothetical genotypes in two environments, showing (a) no GE, (b) 'quantitative' GE (without reversal of ranks of genotypes), and (c) 'qualitative' GE (with rank reversal) (from Romagosa and Fox, 1993).

8.2.1 Case studies

A common approach to studying GE and adaptation is empirical and statistical, and involves describing genotypic responses, usually measured in terms of yield, across a sample of environmental conditions. The contrasting analytical approach characterizes both environments and genotypic responses in terms of biotic and abiotic factors. In practice, most breeding programmes incorporate elements of both empirical and analytical approaches.

Contrasting case studies are described to give some practical context and motivation for studies of GE, to highlight different circumstances and mechanisms involved in crop adaptation and to indicate how strategies and solutions may evolve. Often, after exhausting variation for highly heritable characters, breeding becomes largely empirical and oriented to yield. Genetic progress for unrecognized stresses is, however, often achieved unwittingly by empirical selection within segregating populations. Subsequent recognition of major stresses may result in trait-oriented breeding, with manipulation of the selection environment – by inoculation with diseases for example – to increase the heritability of the traits involved.

For over a quarter of a century, two closely related wheats, 'Gabo' and then 'Gamenya' from eastern Australia, were the mainstay of Western Australian production. 'Gamenya', released in 1960, dominated production until the mid 80s. The regional testing programme was commenced in 1966, following the decision to expand cereal production, and through the 60s, 70s and early 80s the target of Western Australian wheat breeders was to better 'Gamenya'. There appeared to be little heritable variation upon which to select, so a strategy to increase the chance of success by sowing more plots was employed. In 1969, 700 plots per team member were grown and 15 varieties were released between 1974 and 1979. From 1979 to 1984 the total number of releases increased to 44. 'Gamenya' was finally beaten by a series of newer genotypes whose more specific adaptation resulted from combining high selection intensity, using accumulated data to redefine recommendation domains (particularly with respect to the length of growing cycles) and exploiting exotic germplasm with high yield potential. The number of plots per team member stabilized at around 9000–10 000 per year.

Another example emphasizes the plastic and conditional nature of the concept of a domain. The same wheat line was released as 'Antizana' in the Ecuadorian highlands in 1978 and as 'Wilgoyne' in 1989 for the drier parts of Western Australia, two extremely different areas both in terms of available moisture and plant pathogens. It is hypothesized that this wheat's tolerance to zinc deficient soils may give it an advantage in both places.

Drought resistance *per se* in spring wheat is elusive, and photosynthetic efficiency appears to be a major determinant of a genotype's merit across a range of high to low yielding environments. There is a spillover from the genetic gains achieved in highly productive situations to the most stressed. This is consistent with the conclusion of DeLacy, Fox and Corbett (1994) that there was little genotype by moisture availability interaction in international spring wheat yield trials.

In contrast to wheat, improved drought resistance in maize is related to shortening the interval between anthesis and silking (ASI) and there is a spillover in the opposite direction from the wheat example, in that reductions in ASI, for which selection is most efficient under stress, may increase the yield in high yielding agriculture. These gross differences in selection strategies between wheat and maize are explained by the fact that the male and female inflorescences in wheat are contained and synchronized within the glumes while they are separated on the maize plant. Apical dominance and the proximity of the male inflorescence to the most photosynthetically active leaves is thought to favour pollen dissemination at the expense of seed production in wild maize and its progenitors (Bolaños and Edmeades, 1993). Apart from synchronizing flowering, selection for ASI has also contributed to increased yields in the domesticated maize crop by redirecting photosynthate to the female inflorescence, resulting in an increase in the harvest index under a wide range of moisture conditions.

8.2.2 Wide versus narrow adaptation and selection ability of locations

Breeders aim to cover a representative sample of spatial and temporal variation. Sometimes a breeder's selection environments in one year may have little relation to those experienced in the next (Fox, Rosielle and Boyd, 1985). The sampling problem, associated with yearly variation, suggests testing for many crop cycles. However, breeders often substitute temporal variation with spatial variation, assuming that testing over a wide geographic range can ensure a parallel degree of temporal buffering capacity in their germplasm.

Related to wide adaptation is the question of breeding locations: can selection under optimum high-input environments identify genotypes adapted to more stressed environments? Any generalizations are risky but an understanding of the biology of the crop species may help in the choice of strategy. In the example of wheat and maize described, the flowering structures of the species are major reasons for using different selection strategies.

Genotypic performance is influenced by the choice of locations for early generation selection (Campbell and Lafever, 1977), particularly in production areas with variable levels of abiotic stress.

For breeders, environmental diversity for field crops represents a 'double-edged sword'. The diversity complicates breeding and testing of improved genotypes with adequate adaptation, but also permits identification of extreme environmental conditions that guarantee selection pressure from important stresses.

8.2.3 Adaptation to stresses, low inputs and marginal environments

Atlin and Frey (1989) concluded that the need for separate breeding programmes for low- and high-input environments depends on the extent to which the yields in the two types of environments are under separate genetic control, and on the accuracy of selection. Yields in the two environments must be controlled by different alleles to justify different programmes.

However, such issues may be dynamic as breeders strive to incorporate new alleles that are superior in both situations. Apart from theoretical considerations, the argument for or against selection in high versus low yielding environments often relies on circumstantial evidence. An exception is a wheat selection study (Ud-Din, Carver and Clutter, 1992) which showed that the optimum selection index for stressed environments was a weighted mean yield over stressed and unstressed trials with weights in the ratio of 1:2. Ceccarelli (1994), however, favoured and demonstrated the benefit of direct selection under conditions similar to the target environment.

8.2.4 Long-term relationships among locations and among genotypes

Fox and Rosielle (1982b) suggested using a reference set of diverse genotypes to handle unpredictable seasonal variation. A breeder would define a long-term target environment using relative yields or rankings from a reference set of genotypes grown over the years. The net increase in plots would generally not be significant and the reference genotypes would be grown to determine the proximity of the specific location–year combination to the long-term target, in terms of GE. Then results from each location in a year would be weighted in accordance with the location's representativeness or proximity to the target.

Probe genotypes, near-isogenic pairs with differential response to known environmental stimuli, provide a more qualitative diagnostic approach to typing environments, particularly for factors that are not easily observed. A marked divergence in the performance of a pair in a test environment indicates a high probability of a stress, such as a micronutrient imbalance in the soil or the presence of a soil-borne pathogen. Chapter 3 discusses positive feedback between characterization of genotypes and characterization of environments. By judicious selection of data, genotypes which were not specifically developed as probes can also indicate particular stresses.

The alternative approaches of wide testing of advanced lines and exposure to a few 'key locations' with defined stresses can be considered as complementary, with the balance between the two depending upon how well major stresses are defined.

8.3 UNIVARIATE DATA ANALYSIS AND EXPLORATION OF GENOTYPE PERFORMANCE

GE is often best examined through a combination of techniques and Fig. 8.2 provides an analytical protocol that could be followed with trial data. The analysis may be iterative; for example, if subsets of environments are formed, the generation of stability parameters should be repeated within groups. In Fig. 8.2, we have indicated the appropriate sections alongside each analysis. In this section we consider the analysis of variance, regression analysis, risk assessment and ranking methods. Section 8.4 covers multivariate methods.

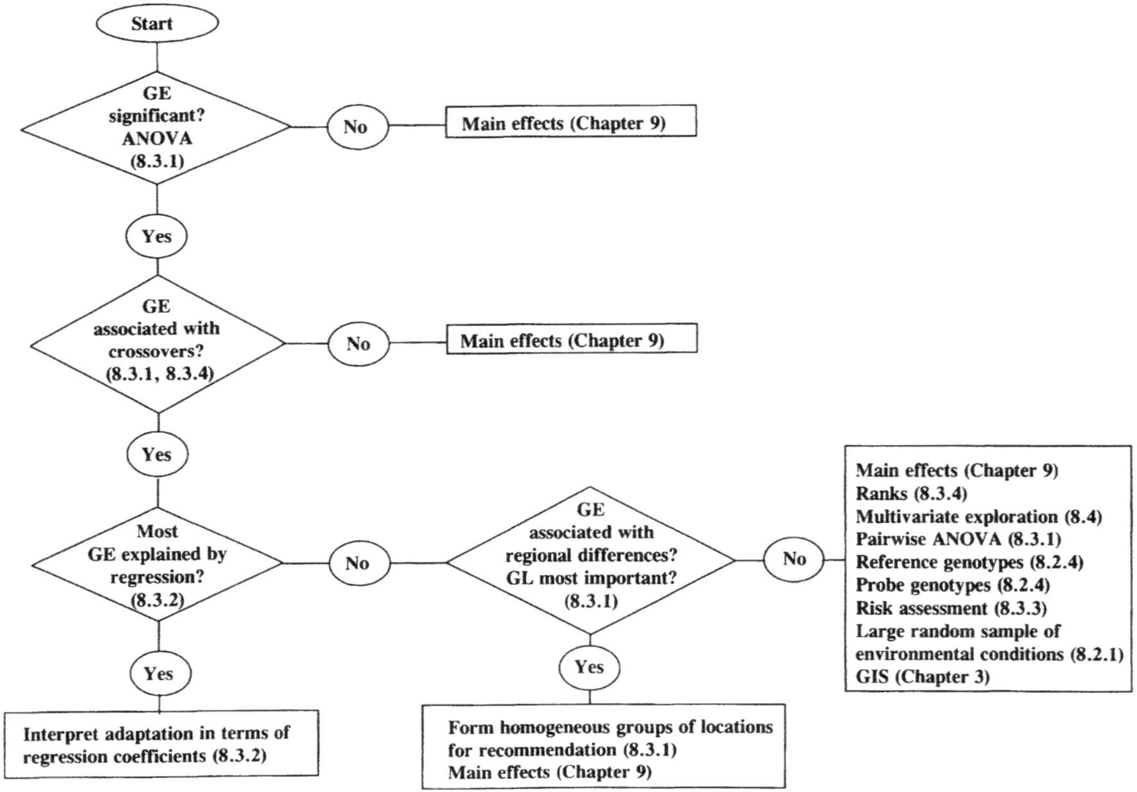

Fig. 8.2. An analytical protocol for GE data exploration.

8.3.1 Analysis of variance

The mean yield Y_{ij} of the ith genotype in the jth environment can be expressed as

$$Y_{ij} = \mu + g_i + e_j + ge_{ij} + \varepsilon_{ij} \quad (8.1)$$

where μ is the general mean, g_i, e_j and ge_{ij} are the effects of genotypes, environments and genotype × environment interactions, respectively, and ε_{ij} is the average random error. It is assumed that ε_{ij} has an independent Normal $(0,\sigma^2/r)$ distribution, where σ^2 is the within-environment error variance (assumed homogeneous) and r is the number of replicates in an environment.

Analysis of variance estimates the GE variance component and tests whether it is significantly different from zero. DeLacy, Eisemann and Cooper (1990), in a synthesis from more than 20 species, highlighted the ubiquitous nature of GE for yield. Talbot (1984) analysed the yields of 15 species over several locations and years and showed how GE could be split into components for G · L, G · Y and G · L · Y (see also Chapter 10). However, GE variance components do not allow a detailed exploration of the patterns of interaction.

Cooper and DeLacy (1994) partitioned the GE variance component into a term for the heterogeneity of genetic variance among environments and another for the lack of genetic correlation. Intuitively, the second term would seem to be related to rank changes of genotypes [although Fig. 8.1(c) shows the rank change with correlation and, conversely, uncorrelated responses do not always involve crossovers]. The pooled genetic correlation among all of the environments can be estimated (Dickerson, 1962) as the intra-class correlation from the variance components of the combined analysis of variance and provides complementary information on heritabilities. Where GE accounts for a large proportion of the variation among genotypes, the individual genetic correlations could be investigated, for example when examining specific pairs of locations within a breeding programme, rather than the pooled genetic correlation. Cooper and DeLacy (1994) showed that as the magnitude of the GE component of variance increases relative to the genotypic component, both heritability and response to selection are reduced.

Often a breeder's assessment of a genotype is based on comparison with one or more standards. Lin and Binns (1985) suggested a pairwise analysis of variance between individual test genotypes and standard check varieties to detect which genotypes show the same adaptation pattern as checks in regional yield trials. Genotypes that showed no significant GE with a check were identified and those with significantly higher yields across locations were recommended.

8.3.2 Linear regression analysis

Linear regression provides a conceptual model for genotypic stability and is often known as the Finlay and Wilkinson (1963) approach, after its main popularizers. The slope, β_i, of the regression of an individual genotype's values against an environmental index estimates genotypic stability. The index is usually estimated as the mean of all genotypes in a trial.

The GE term from Equation 8.1 may be partitioned between the heterogeneity of regressions (differences in slopes, β_i) and deviations from linear regressions. The main limitation to regression methodology and its extensions (Crossa, 1990) is biological, specifically the juxtaposition of ecologically different environments with similar mean yields on the *x*-axis can mask linear relationships for individual factors, for example, frost (Knight, 1970). If the fraction of the GE sum of squares explained by regression is low, regression coefficients should be interpreted with caution. A genotype with a unique tolerance to a stress such as a soil-borne nematode may show larger deviations from regression than average genotypes, which numerically dominate the trial results and are thus likely to define the environmental index.

8.3.3 Risk assessment

Particularly in subsistence agricultural systems, yield *per se* may be less important than reaching a certain yield level. Safety-first selection (Roy, 1952) minimizes the chance of falling below a specified yield level in a recommendation domain. How can a breeder decide between two genotypes, one of which has a lower mean yield and a smaller variance than the other? The safety-first concept selects the genotype with the smallest chance of producing a yield less than some minimally acceptable value d (Eskridge, Byrne and Crossa, 1991). A plant breeder follows the safety-first concept by selecting genotypes that minimize the probability of failure.

Assume that yields (Y_i) for the ith genotype are normally distributed with mean μ_i and variance σ_i^2. Genotypes that minimize the probability that Y_i is less than d are preferred. Now

$$P(Y_i \le d) = P\left[\left(\frac{Y_i - \mu_i}{\sigma_i}\right) \le \left(\frac{d - \mu_i}{\sigma_i}\right)\right] = F\left(\frac{d - \mu_i}{\sigma_i}\right) = 1 - F\left(\frac{\mu_i - d}{\sigma_i}\right)$$

where F is the standard cumulative normal distribution function, and so the genotype which minimizes $P(Y_i \le d)$ also maximizes $F[(\mu_i - d)/\sigma_i]$. Since $F(\cdot)$ is an increasing one-to-one function, choosing the genotype with the largest value of $F[(\mu_i - d)/\sigma_i]$ is the same as choosing the genotype with maximum value of $(\mu_i - d)/\sigma_i$. Thus, if $\bar{Y}_i$ and s_i^2 are the sample mean and variance of genotype i across the environments, the genotype that minimizes the probability of $Y_i < d$ is the one with the largest value of $(\bar{Y}_i - d)/s_i^2$.

Values of d should be selected on the socioeconomic circumstances faced by the target farmer. For a subsistence farmer, d may be the minimum yield necessary to feed the family, whereas for a commercial farmer the value of d is that required to meet variable costs of production. A farmer with access to credit could alternatively try to maximize his long-term production while accepting crop failures in difficult years.

The relationship of d with the probability of failure of a genotype is shown in Fig. 8.3. The distributions of four hypothetical genotypes are depicted in

Fig. 8.3(a); genotypes *G2* and *G3* have the same mean (and median) but different variances, whereas genotypes *G1*, *G2* and *G4* have different means (and medians) but the same variance. In Fig. 8.3(b) values of *d* are plotted against the hypothetical cumulative normal distribution functions of the four genotypes (proportion of trials for which the genotype yields less than *d*). The values of *d* corresponding to *M1*, *M2*, *M3* and *M4* represent the median of genotypes *G1*, *G2*, *G3* and *G4*, respectively. For example, for *G1*, 50% of the trials yielded less than $d = M1$, whereas for *G4* the median is higher $d = M4$. For subsistence farmers with a value of *d* such as d_s in Fig. 8.3(b), *G4* is the best genotype because it has (i) the lowest probability of producing yields less than d_s; (ii) the largest median (*M4*); (iii) smaller variance than *G3*; and (iv) large lower quantiles (e.g. the 10% quantile corresponds to the largest value of *d*). For the same value, $d = d_s$, the genotype with the highest risk of yielding below d_s is *G1*. Furthermore, *G1* has a high probability (> 0.9) of yielding below $d = M2$. For commercial farmers, the appropriate *d* value may be larger than d_s (say d_c). In this situation, genotype *G4* would still be the best genotype because of the small proportion of trials yielding below d_c. For values of $d < M2$ (or *M3*), *G2* is a better choice than *G3*; however, the situation is reversed for values of $d > M2$ (or *M3*). For $d = M4$, the first choice would be *G4*, followed in order by *G3*, *G2* and *G1*.

In general, a plant breeder is concerned with developing a genotype with (i) better mean performance than accepted genotypes; (ii) reasonable stability across a wide range of environments; and (iii) low probability of failing to achieve a minimum acceptable yield.

When selecting in the presence of GE, the breeder must weight the importance of a genotype's stability relative to its average performance over environments. Eskridge (1990) and Eskridge, Byrne and Crossa (1991) developed safety-first selection indices based on different stability approaches and concluded that if (i) the means and stability parameters of genotypes differ; (ii) GE is large; and (iii) poor yield has severe consequences, then a safety-first index is useful to weight the importance of stability relative to mean yield.

Although the probability of an individual genotype failing to achieve a minimum acceptable yield is important, the concern of most plant breeders is the development of a genotype that performs better than the standard local check. Eskridge and Mum (1992) used a safety-first screening approach based on selecting test genotypes that outperformed the local checks with a certain probability. For an international maize yield trial, they found that this approach is useful when the costs of introducing a genotype that fails to outperform the local adapted checks are high.

Stochastic dominance (Anderson, 1974) ranks genotypes according to suitability for risk-averse farmers. Suppose that the probability distribution of yields for the *i*th genotype is $f_i(d)$ and its cumulative distribution function is $F_i(d)$. If all the values of the cumulative yield distribution of genotype *i* are greater than those of genotype *k*, that is $P(Y_k \le d) \le P(Y_i \le d)$ for all *d* or $F_k(d) \le F_i(d)$, then f_k is said to have first-degree stochastic dominance over f_i, i.e. genotype *k* is more risk-efficient than *i* [Fig. 8.4(a)]; f_k has second-degree stochastic dominance over f_i if $P(Y_k \le d) \le P(Y_i \le d)$ for $d < d_0$ and the area A, in Fig. 8.4(b), is greater than the area B.

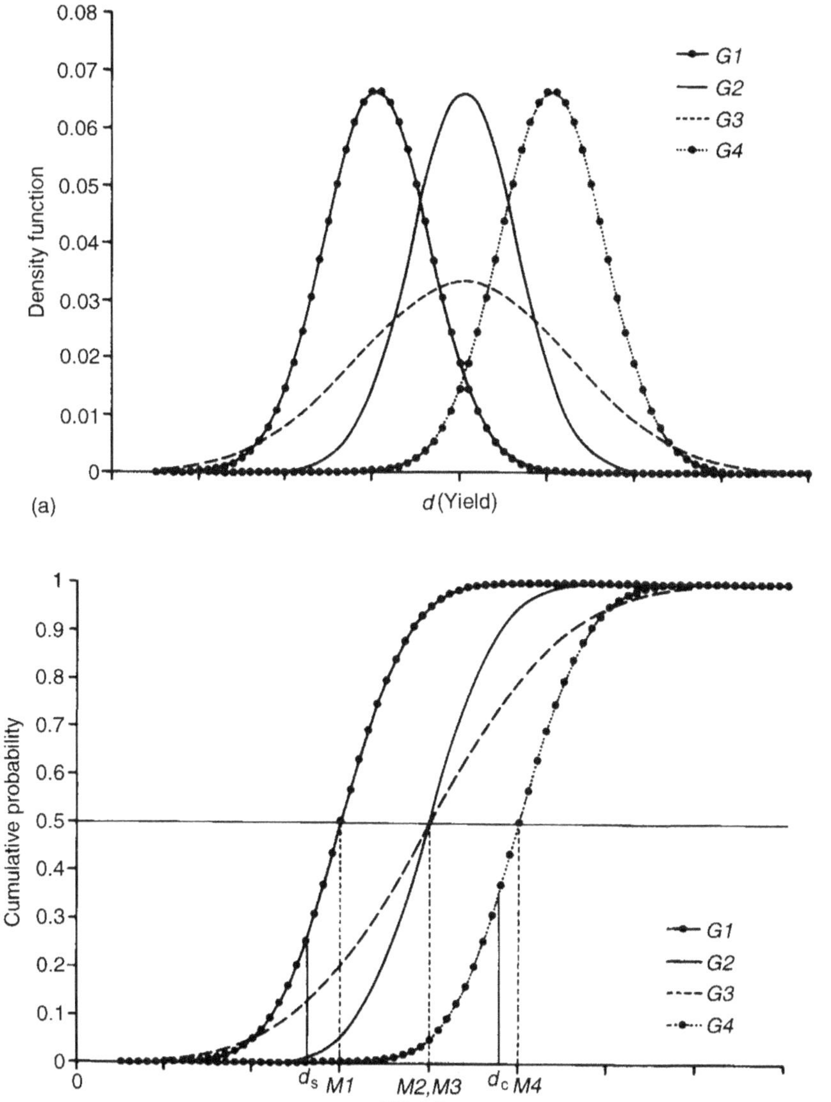

Fig. 8.3. Risk assessment for four hypothetical genotypes, *G1*, ... ,*G4*: (a) yield distributions and (b) cumulative probabilities of achieving the given yield. *M1*, ... ,*M4* are the medians of the distributions of genotype yield; d_s and d_c represent the minimum yield acceptable for subsistence and commercial farmers.

For Normally distributed genotype yields (e.g. Fig. 8.3), the condition for first-degree stochastic dominance of genotype k over genotype i is that $\mu_i < \mu_k$ and $\sigma_i^2 \geq \sigma_k^2$. This is the criterion for risk efficiency proposed by Binswanger and Barah (1980). However, if the Normality assumption is not reasonable

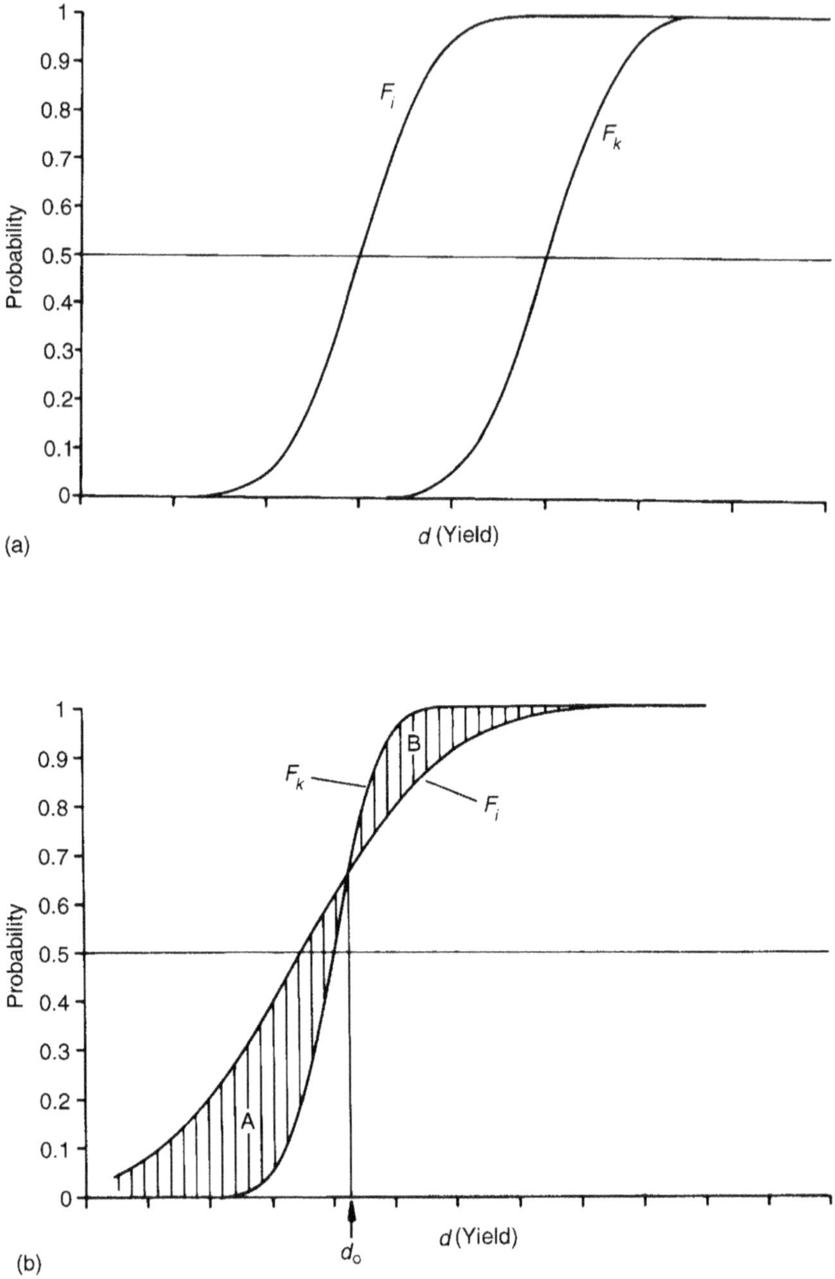

Fig. 8.4. Cumulative probability distributions F_i, F_k, for genotypes i and k showing (a) first-degree stochastic dominance of genotype k over genotype i, (b) second-degree stochastic dominance, and (c) where first- and second-degree stochastic dominance do not hold, but the safety-first criterion selects genotype k below d_o.

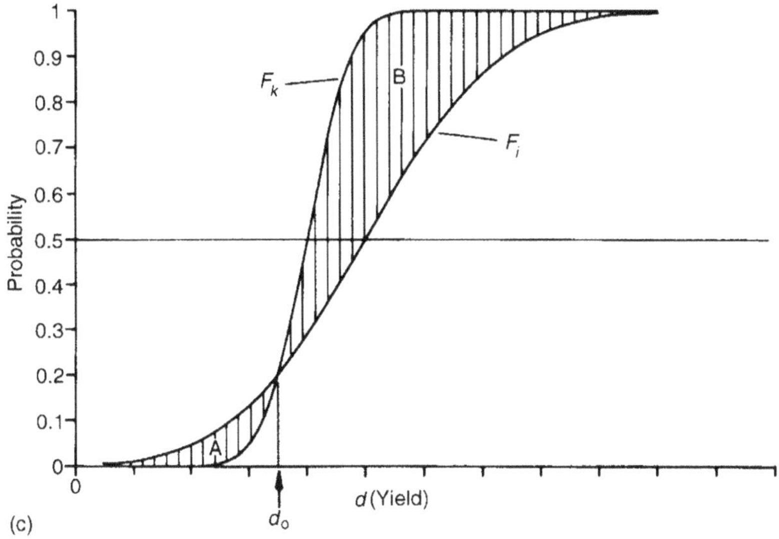

Fig. 8.4. (continued)

we can compare the empirical cumulative distributions, provided the number of environments is sufficient.

First-degree stochastic dominance is an extension of the safety-first procedure of Eskridge (1990), Eskridge, Byrne and Crossa (1991) and Eskridge and Mum (1992) for all values of d. However, when first- and second-degree stochastic dominance do not hold, safety-first procedures can identify the optimal choices of genotypes for given d values. For the example in Fig. 8.4(a), $P(Y_k \leq d) \leq P(Y_i \leq d)$ for all values of d, first-degree stochastic dominance exists and genotype k is the better genotype. For the example in Fig. 8.4(b), second-degree stochastic dominance indicates that for all of the values of d up to d_o, $P(Y_k \leq d) \leq P(Y_i \leq d)$. Thus the better genotype for all d up to d_o is genotype k; a similar result is obtained with the safety-first criterion. Figure 8.4(c) shows that first- and second-degree stochastic dominance do not hold because area A is smaller than area B; however, safety-first shows that $P(Y_k \leq d) \leq P(Y_i \leq d)$ for all d up to the point d_o. Thus, for all $d < d_o$, genotype k is still preferred. However, at lower values of d_o the accuracy with which the crossover point is determined will generally be lower than at higher yields levels.

Using stochastic dominance, Menz (1980) found that newer entries in international wheat trials were higher yielding and more risk efficient than older ones. Wheats with high risk efficiency and high mean yield performed well across all environments.

8.3.4 Ranking methods

Analysis of ranks addresses the problem of cross-over interactions directly. Advantages of ranking methods include (i) freedom from assumptions concerning additivity of main effects, homogeneity of variances and linear response to increasing environmental yield potential; and (ii) measures of adaptation are not unduly influenced by genotypic performance in extreme environments. Stratified ranking (Fox, Skovmand and Thompson, 1990) evaluates the proportion of environments where any genotype ranks in the top, middle or bottom third of the entries, but takes no account of the errors involved. A genotype usually found in the top third of entries across the environments is relatively well adapted.

Regardless of their yield level, two environments may be considered alike for selection purposes if both rank genotypes similarly. The similarity of two rankings may be estimated by Spearman's rank correlation coefficient. Rank coefficients can be used for grouping environments or genotypes.

8.4 MULTIVARIATE EXPLORATION OF RELATIONSHIPS AMONG ENVIRONMENTS AND AMONG GENOTYPES

The previous section focused on describing genotypes. However, relationships among environments are also critically important in plant breeding, especially in determining domains for selection and recommendation of genotypes.

Chapter 3 described the feedback between environmental characterization and genotypic characterization. Biological understanding of one strengthens the other process. This biological relationship has a strong statistical parallel in the analysis of two-way GE tables: this can be oriented towards the relation among genotypes or among environments, but conclusions about one aspect may be conditional upon features of the other.

8.4.1 Pattern analysis

Multivariate techniques can describe relationships among environments and among genotypes. A genotype's performance can be described in multidimensional space, with each dimension representing a test environment, the coordinates for which are the measured yields (or other traits). Conversely, environments can be considered in multidimensional space, with each dimension a genotype (Fig. 8.5). Yield data from genotype × location tables commonly generated by breeding programmes in each crop cycle can be summarized through multivariate statistics.

Such methods originated in studies of numerical taxonomy and have been increasingly applied to data exploration in plant breeding since the landmark study of the Fourth International Spring Wheat Yield Nursery by Byth, Eisemann and DeLacy (1976). The term pattern analysis (Williams, 1976) describes the parallel use of classification and ordination techniques to represent the essential variation from genotype × environment tables in a few dimensions. Classification techniques, such as cluster analysis, presume discontinuities within data, while principal components analysis (PCA) and other methods of ordination assume samples (genotypes and environments)

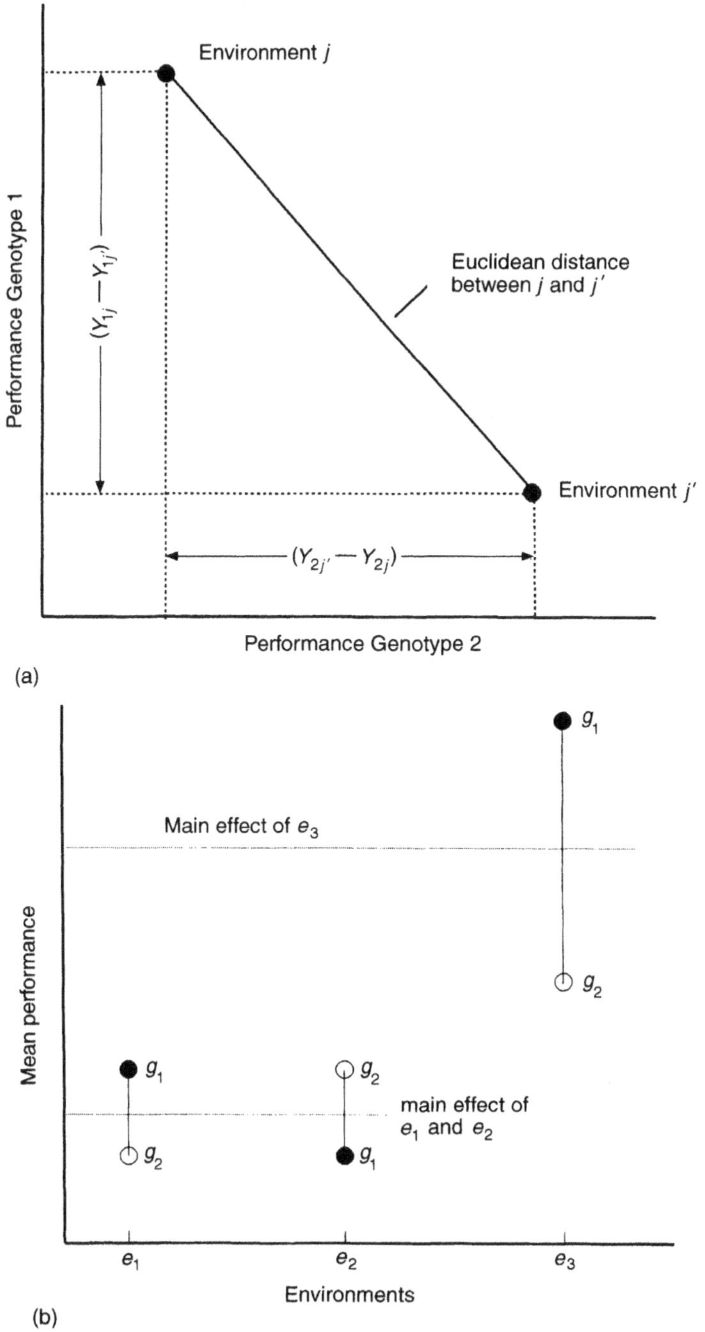

Fig. 8.5. Separation of environments in genotypic space: (a) distance between two environments in two dimensions representing genotypic performance, (b) performance of two genotypes, g_1 and g_2 grown in three environments, e_1, e_2 and e_3, and (c) their distance based on environment-centred data. Note that while e_1 and e_3 rank the genotypes the same way, these environments are widely separated in Euclidean space based on unstandardized data (from Fox and Rosielle, 1982a).

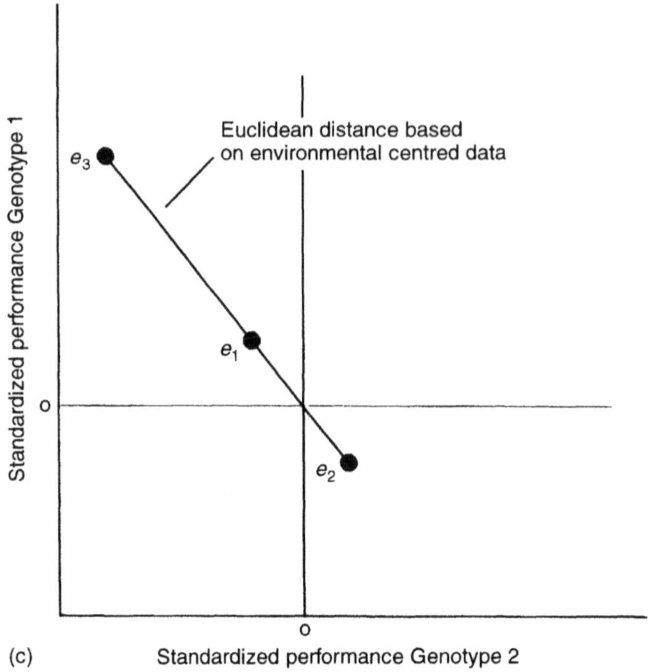

Fig. 8.5. (Continued)

come from a continuous distribution. Thus classification and ordination emphasize slightly different features of data. Pattern analysis is used mainly as a technique for data exploration and makes no *a priori* assumptions, such as linearity or normality, about the form of the genotypic response.

The concept of separation of environments in multidimensional space was explained with two-dimensional analogues by Fox and Rosielle (1982a). Considering the performance of each of the two genotypes as axes, a two-dimensional scattergram of the relationships among the environments is obtained (Fig. 8.5). The distance between two environments is the length of the straight line connecting them and the squared Euclidean distance is the square of this length. This measure is extended to many dimensions or genotypes and is used as a measure of the dissimilarity among environments. Similarity is defined as the complement of dissimilarity and the two are collectively described as proximity measures. Other proximity measures used in pattern analysis include covariances of genotypic yields, product moment correlations and rank correlations. However, ranks do not preserve the relative magnitude of the difference in yield between two consecutively ranked genotypes.

In cluster analysis a fusion strategy is selected to process the proximity matrix. Cooper *et al.* (1993) recommended the use of incremental sums of squares (Ward, 1963), an hierarchical clustering routine. Starting from single environment groups, at each fusion the new group formed is the one which minimizes the increase in the within-groups sums of squares. (For any specified number of groups, it is usually possible to find a non-hierarchical grouping, which gives a smaller within-group sum of squares (Digby and Kempton, 1987), but at a greater computational cost.) Fusion by incremental sums of squares tends to form groups of uniform size, avoiding the problems of biological interpretation associated with single-member groups. To a degree, groups of environments and genotypes formed may be algorithm-dependent, but different patterns from different procedures and starting points may indicate that there is no one correct solution in a biological sense.

Multivariate methods often present most of the total variation in a few dimensions, for instance, in a dendrogram from cluster analysis or scattergram from PCA. The two displays may be combined by superimposing the groups obtained from clustering on the PCA scattergram. This combined display summarizes the essential information in the proximity matrix in a readily interpreted form.

Multivariate exploration of data can also be used to group genotypes that perform similarly across environments. The genotypic relationships obtained from wide testing have shown a repeatability from year to year and thus, by association with known standard varieties, a quick assessment of adaptation of advanced genotypes is obtained.

When considering the patterns of the relationships among environments (usually locations or location–year combinations) for breeding, environment main effects are removed so that the grouping is determined more by GE effects than by production level (Fox and Rosielle, 1982a). Further, Cooper *et al.* (1993) argued that the opportunities for exploiting indirect selection among the environments are maximized by standardizing the data such that all of the environments have the same mean and variance.

In contrast to the repeatability of groupings of genotypes, repeatability of location groupings among the years is low. Inter-location relationships based on one year's data are often dominated by unpredictable weather patterns and not easily interpreted. Pattern analysis of data from single years thus provides rapid assessment of genotypic performance and a 'snap shot' of oscillating inter-location relationships. Lack of a sufficient number of common genotypes has limited analyses across the years. Peterson and Pfeiffer (1989) overcame this problem by constructing a long-term correlation matrix of locations using 17 years of yield data from the International Winter Wheat Production Nurseries. They reasoned that the phenotypic correlation between any pair of locations in a given year measures biological similarity and, even though the genotypic composition changes with time, pooling correlations over years would provide a long-term average measure of the relatedness of locations, like superimposing a series of snap shots to produce the long-term picture.

Kroonenberg and Basford (1989) extended the application of PCA beyond two-way data tables to a further dimension, with genotypes by environments tables considered for more than one trait or attribute. Basford *et al.* (1990) provided a parallel development for cluster analysis and Crossa *et al.* (1995) used both the three-way methodologies to characterize maize germplasm from Mexico. These methods can be considered multivariate extensions of earlier work on selection indices that attempted to combine multi-attribute information into a score. Multi-attribute analyses have not explained yield in terms of other measured variables. They may have the greatest potential for application in assessing traits that are truly multivariate in nature: milling and baking data for wheat, potato yield grades, ratoon yield for sequential harvest or time series analysis of disease progress.

8.4.2 Additive Main effects and Multiplicative Interaction model

The Additive Main effects and Multiplicative Interaction (AMMI) model is an analytical tool to interpret GE (Gauch and Zobel, 1988; Crossa *et al.*, 1991). AMMI is a form of principal components analysis and an ordination method. It has, however, developed its own terminology and interpretation and thus warrants a separate section. The complete AMMI model is

$$Y_{ij} = \mu + g_i + e_j + \sum_{k=1}^{t} \lambda_k \alpha_{ik} \gamma_{jk} + \varepsilon_{ij} \quad (8.2)$$

where Y_{ij} is the cell mean of the ith genotype in the jth environment, μ is the overall mean, g_i and e_j are the main effects of the genotypes and environments, respectively, and ε_{ij} is the experimental error. Residuals of the (ij)th cell $Z_{ij} = Y_{ij} - Y_{i.} - Y_{.j} + Y_{..}$ form a matrix **Z**. The least squares estimates of the AMMI parameters, λ_k, α_{ik}, γ_{jk}, are the eigenvalues and eigenvectors of the matrix **Z′Z**.

AMMI generates a family of models. AMMI0 uses the additive genotypic and environmental means to describe the data matrix and thus ranks genotypes identically at each environment, ignoring GE. The second model, AMMI1 ($Y_{ij} = \mu + g_i + e_j + \lambda_1 \alpha_{i1} \gamma_{j1} + \varepsilon_{ij}$), considers the main effects as well as one principal component axis (PCA1) to interpret the residual matrix. AMMI2 ($Y_{ij} = \mu + g_i + e_j + \lambda_1 \alpha_{i1} \gamma_{j1} + \lambda_2 \alpha_{i2} \gamma_{j2} + \varepsilon_{ij}$) considers the main effects plus two axes, PCA1 and PCA2, for non-additive variation. Subsequent models consider an additional PCA. The higher order multiplicative components that are not significant can be ignored, giving a 'reduced' model.

Gauch and Zobel (1988) considered two methods for choosing an appropriate AMMI model for a GE data set. In the descriptive or 'postdictive' procedure, models are fitted with increasing numbers of axes and the significance of each additional multiplicative term is tested using an approximate F-statistic (Gollob, 1968). The cross-validation or 'predictive' procedure relies on several replicates being available for each GE combination, one of which is used to validate the AMMI model fitted to the other replicates. The number of multiplicative terms that best predicts the validation set is selected, and this model is refitted using the full data set based on all of the replicates. The advantage of the cross-validation procedure is that separate data are used for fitting and testing the models. Gauch and Zobel (1988) found that with the standard descriptive procedure there was a danger of choosing a large number of axes and of over-fitting the data. This means that the model incorporates data noise, which is not replicated in the validation set. In particular, the maximal AMMI model, equivalent to the cell means, was found to be a poor predictor of the validation set. In general, the cross-validation procedure shows that GE data sets are best described by AMMI models with one or two multiplicative terms (Gauch and Zobel, 1988).

Gauch (1992) advocated using the genotype means from the selected AMMI model, rather than from the additive model, to predict overall performance in a region. This is controversial and assumes that the validation based on the additional replicates of a set of environments is relevant for predicting the performance in new environments (Kempton, 1994b). On the other hand, we suggest that the interactions within well-formed recommendation domains will be unlikely to have useful predictive value. For this reason, the methods of prediction advocated in Chapter 9 are based on the additive model.

Crossa *et al.* (1991) pointed out the risk that relying solely on the AMMI model to describe GE may lead to important interactions from a single environment with a unique stress being overlooked. Large residuals from the AMMI model should therefore be carefully monitored to indicate how successfully data are modelled. Another problem related to AMMI analysis is the phenomenon of a homogeneous subset of data indicating a more complex pattern than the complete set. This apparent paradox suggests that the concept of noise extraction by AMMI analysis requires more investigation (Romagosa and Fox, 1993).

When AMMI1 provides a good description of the data, the main effects and PCA1 scores for genotypes and environments are plotted on the same diagram (see for example, Romagosa and Fox, 1993, Fig. 22.4), facilitating inference about the specific interactions of individual genotypes and environments by using the sign and magnitude of PCA1 values. Any genotype with a PCA1 value close to zero shows general adaptation to the tested environments. Its response pattern across the environment parallels the mean of all the genotypes in the trial. A large genotypic PCA1 score reflects more specific adaptation to environments with PCA1 scores of the same sign. This display is a particular form of biplot, proposed for presenting GE data by Kempton (1984b).

8.4.3 Biplots

Here we follow the biplot procedure of Kempton (1984b) to provide an alternative to the AMMI presentation (Romagosa and Fox, 1993) of Spanish triticale data of Royo, Rodriguez and Romagosa (1993), using an AMMI2 rather than AMMI1 model.

Figure 8.6 plots the genotypes and environments on the first and second PCA axes, and the vectors originating at the origin are shown for each environment. The length of an individual vector indicates the amount of GE variation modelled for that environment. The first axis separates Orense, Coruña and Salamanca (all acid soils) from other locations and triticales (*C1–C6*) with seven rye chromosomes from those with six. Interaction effects of the genotypes in an environment are approximated by their perpendicular projections onto the environmental vector, or its extension. For example, in Fig. 8.6 the

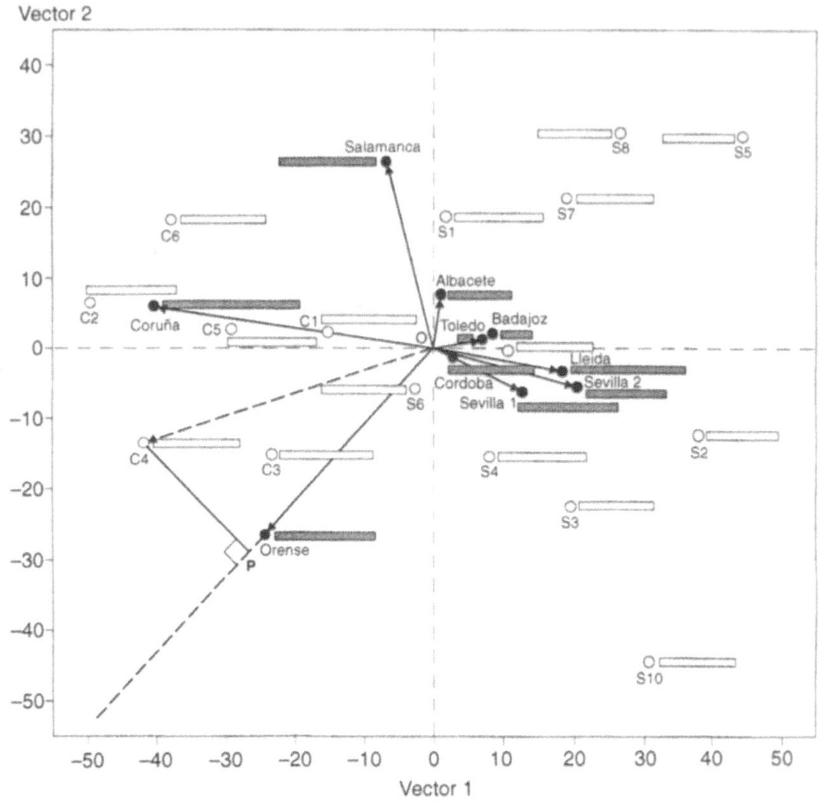

Fig. 8.6. Biplot of Spanish triticale data (Royo, Rodriguez and Romagosa, 1993) Six complete triticales are designated *C1* to *C6* and the substituted triticales are *S1* to *S10*. These 16 genotypes are indicated with an open circle. The locations are indicated by dark circles. Sevilla was represented by two trials and the other locations by one. The lengths of the horizontal bars, open for genotypes and shaded for locations, are proportional to grain yield.

interaction effect for genotype *C4* in Orense is proportional to the length OP. The AMMI2 model predicts that *C4* has the highest positive interaction in Orense, followed by *C2* and *C3*. (The interaction effect is negative when a projection intersects a negative extension of a vector.) It is essential that both axes of the biplot are plotted on the same physical scale – otherwise distortion of the lengths of vectors and angles between them will occur.

One of the useful features of the AMMI1 presentation is that the size of the environmental variation compared with the genetic variation is highlighted by the main effects axis. To incorporate these main effects into the AMMI2 biplot of Fig. 8.6, we represent each genotype and environment with a bar whose length is proportional to its mean yield.

Environmental vectors at right angles in a biplot represent the environments for which the correlation of the GE effects is close to zero. If the angle between vectors is less than 90°, the correlation is positive; and if the angle is greater than 90°, negative. Nearly parallel vectors with the same direction (the two Sevilla trials in Fig. 8.6) indicate high positive correlations, and with opposite directions (Sevilla and Coruña) indicate strong negative correlations for GE. These diagnostic features can be a powerful aid for forming recommendation domains.

8.5 INFORMATION EXTRACTION VERSUS COMPUTATIONAL COMPLEXITY

In this summary we consider the GE methods of the chapter in terms of the information extracted and their computational complexity. Figure 8.7 suggests a positive relation between information and complexity. Although computational complexity can be influenced by the availability of appropriate, user-friendly software for the analyses, we suggest that underlying statistical complexity often goes hand in hand with the complexity of the biological interpretation.

Figure 8.7 emphasises our view of the benefits of ranking methods, in

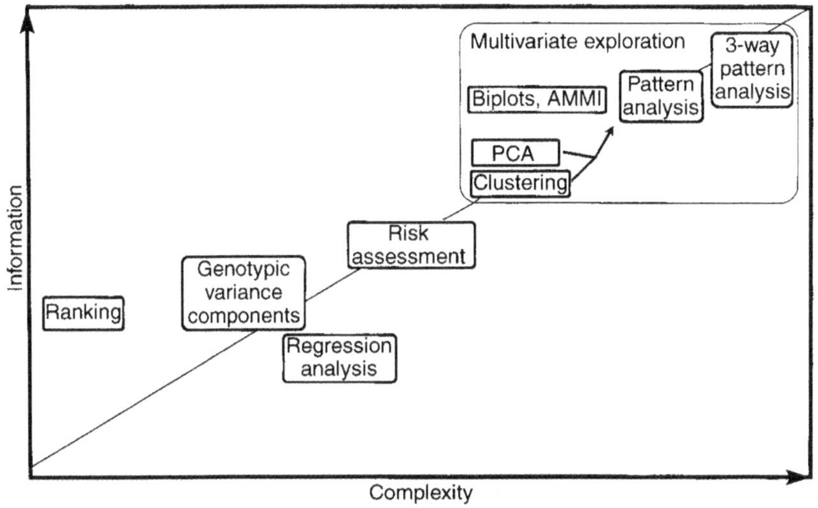

Fig. 8.7. Complexity versus information content of analytical methods (modified from Souza, Myers and Scully, 1993).

terms of simplicity and information extraction, to a plant breeding programme without recourse to major computational power. (We acknowledge that this is a breeder's perspective: a population geneticist might consider that we have underestimated information from genetic variance components.) At the same time, we hope that software development increasingly will put multivariate exploration in the breeders' hands. There are, however, no easy roads towards biological interpretation, although the advances in data management and querying outlined in Chapter 3 may be powerful tools. In the field of GE, it is essential that statistical expertise should be matched with parallel biological understanding of the crop involved.

9

Analysis of series of variety trials

H.D. PATTERSON

9.1 INTRODUCTION

This chapter describes the statistical analysis of series of trials aimed at predicting the future performance of new varieties in a specified region or country (a recommendation domain in the terminology of Chapter 8). The methods have been developed to meet the requirements of official variety testing in the UK (Patterson and Silvey, 1980) and are based on principles established by Fisher (1935) and Yates and Cochran (1938). Varieties entering the trials are distinct entities that have passed through the breeding stage and are now ready for marketing.

Experimenters have long recognized that a single trial provides an inadequate test of the value of a new variety. Trials are needed in several years to provide a range of seasonal conditions and at several locations to sample the main growing areas of the crop.

Analysis is in two stages. First, we analyse each trial individually to give local variety means and estimates of the experimental error. Second, we assemble the local variety means in a summary table classified by varieties, locations and years. Analysis of this table provides estimates of the predicted variety means for the sampled population and the sampling errors of these estimates.

Summary tables may be complete, with every variety in every trial, or incomplete. When the table is complete the required variety means are simple averages and the analysis of variance provides mean squares for estimating components of error variance and sampling errors.

Immer, Hayes and Powers (1934) described one of the first examples to use the analysis of variance. Their series tested ten varieties of barley at six experimental stations in Minnesota, USA, over two years. Each variety appeared in each trial, so that the summary table was complete and the analysis of variance applicable. Fisher (1935, Section 30) used yields of five of the varieties as a practical example in his book *The Design of Experiments*. Section 65 also includes general advice on the use of interactions as error. Yates and Cochran

Statistical Methods for Plant Variety Evaluation. Edited by R.A. Kempton and P.N. Fox. Published in 1997 by Chapman & Hall, London. ISBN 0 412 54750 3.

(1938) re-analysed Fisher's subset of Immer's data and found that one variety was more responsive than others to differences between locations. In consequence, the basic analysis of variance underestimates the errors of contrasts involving this variety and overestimates the errors of other contrasts. They solved this problem by modifying the analysis of variance and calculating additional components of variance.

This chapter extends these classical methods to the analysis of incomplete as well as complete summary tables. Official series of UK trials almost invariably produce incomplete tables (Silvey, 1978) and this complicates the analysis. Simple averages are not adequate for estimating the variety means because differences between them are affected by the generally much larger differences between trials. Moreover, the analysis of variance no longer provides a convenient method for separating components of error variance.

Until recently, incomplete summary tables have usually been analysed either by restricting the analysis to comparisons between each candidate variety and one or more control varieties or by fitting a linear model with a single error term. These methods have worked reasonably well in practice but are not efficient. Generalization of Yates and Cochran's (1938) analysis consists of fitting linear models with several error terms. The residual maximum likelihood algorithm (REML) described by Patterson and Thompson (1971) is suitable for this purpose.

Section 9.3 outlines the main features of REML. In Section 9.4 we show that applying REML with a suitable model to Immer's data gives the same results as Yates and Cochran's analysis. The remaining sections give examples of the analysis of incomplete summary tables using REML and more traditional methods.

9.2 NOTATION AND TERMINOLOGY

9.2.1 Factors

Throughout this chapter we use V, L and Y to represent varieties, locations and years respectively. We use the word variety (V) rather than genotype (G) as our primary concern is with testing finished varieties with individual identities. Additional factors are sometimes needed. In the examples, these include W, a factor representing contrasting groups of varieties (Sections 9.4 and 9.6), a second locations parameter M (Section 9.8) and a factor S allowing for changes in the method of seed production (Section 9.7).

9.2.2 Tables

The formula V × L defines the structure of a two-way table classified by varieties and locations and V × Y the structure of a table classified by varieties and years.

We consider two kinds of three-way table classified by varieties, locations and years. In one, the same locations are used each year and the formula is

$V \times L \times Y$. In the other, the locations are chosen afresh each year and the formula is $V \times (Y / L)$. Formulae of this type have been discussed by Nelder (1965).

9.2.3 Models

We also need formulae for models that are to be fitted to incomplete summary tables. Formulae are already available that specify the terms of an analysis of variance (Wilkinson and Rogers, 1973) or a linear model with a single error component (McCullagh and Nelder, 1983). A simple modification suffices to extend these formulae to linear models with multiple errors.

For example, the single list of effects

$$V + L + Y + L \cdot Y + V \cdot L + V \cdot Y$$

specifies the analysis of variance of a balanced $V \times L \times Y$ table. The modification consists of replacing the single list of effects by two lists, one giving the systematic effects of the model and the other the random error terms; these two lists are separated by a colon. For example

$$V + L + Y + L \cdot Y : V \cdot L + V \cdot Y \quad (9.1)$$

represents a model in which there are three error terms for differences between varieties, one based on $V \cdot L$ interaction, one on $V \cdot Y$ interaction and the other on residual error compounded of the three-factor $V \cdot L \cdot Y$ interaction and experimental error. Conventionally, only the first two of these error terms are included in the formula; the third is always implied.

The variances of the error terms are denoted by σ_{VL}^2, σ_{VY}^2 and σ^2 in the usual manner for variances and components of variance; when they need to be described in words they are called the varieties × locations or $V \cdot L$ component of variance, the varieties × years or $V \cdot Y$ component of variance and the $V \cdot L \cdot Y$ component of variance. The last of these is commonly called the units variance. The number of variance parameters is always one more than the number of terms displayed to the right of the colon in the model formula.

We can also regard L, Y and $L \cdot Y$ as error terms, not of varietal differences but of the estimated mean over all varieties. The model then becomes

$$V : L + Y + L \cdot Y + V \cdot L + V \cdot Y \quad (9.2)$$

As a bonus, fitting Equation 9.2 gives more appropriate weight than Equation 9.1 to any information on varietal differences that might be present in contrasts between trials. The amount of additional information is, however, usually small.

9.3 FITCON AND REML

Most major statistical computing programs now provide facilities for fitting linear models with a single error term to classified data using least squares. We refer to them collectively as FITCON because Yates (1933), who played a major role in developing the method, called it the method of fitting constants. In recent years, FITCON has been widely used for variety assessment in the UK (Silvey, 1978).

REML estimates the components of variances by maximizing the likelihood of all contrasts with zero expectation; hence the name residual maximum likelihood and the acronym. Equations of estimation are solved by successive approximations. Variety means are estimated by generalized least squares with weights depending on the estimated variance components.

Versions of the REML algorithm are now available in several programs including GENSTAT and SAS. The latter also allows full maximum likelihood (ML) to be used instead of residual maximum likelihood. ML and REML give different answers. REML is preferred here because it gives the same results as the analysis of variance method when the summary table is complete and balanced; ML does not. The original algorithm (Patterson and Thompson, 1971) was designed for linear models with qualitative covariates, but continuous covariates can be added in most implementations.

At present REML is still being developed. Computer runs are several times longer than FITCON runs, facilities for constructing tables of predicted means are limited and tests of significance difficult to interpret. Nevertheless this is, in my view, the path to follow.

The following procedure is recommended for fitting REML models to summary tables.

1. Always start with a basic model (see the list below).
2. If an estimated component of variance is negative remove the corresponding term from the model and refit.
3. Check for evidence of unusual sensitivity of one or more varieties to changes of environment. If necessary modify the model and refit.

This procedure closely and deliberately follows that adopted by Yates and Cochran (1938) in their analysis of Immer's data (Section 9.4). Negative estimates of the components of variance are not uncommon, particularly in the analysis of three-way tables. Unfortunately, none of the commercially available versions of REML currently allows automatic refitting at stage 2. This hinders routine application of the method.

The usual criteria for model building do not apply when fitting a REML model to a summary table. For example, dropping a real but non-significant $V \cdot L$ interaction could lead to an underestimation of error. The object of the exercise is to predict variety means for the future, not to find the model that best fits or 'explains' past data.

The most useful basic models are as follows.

Table structure	Basic REML model
$V \times L$	$V : L$
$V \times Y$	$V : Y$
$V \times L \times Y$	$V : L + Y + L \cdot Y + V \cdot L + V \cdot Y$
$V \times (Y / L)$	$V : Y + Y \cdot L + V \cdot Y$

The following sections fit these models to data from a number of different trial series, and discuss their appropriateness and the interpretation of results.

9.4 ANALYSIS OF IMMER'S BALANCED SERIES

The data in Table 9.1 are those used by Yates and Cochran (1938) but have been converted to units of t/ha.

Table 9.1. Immer's data: yields of five varieties of barley (t/ha) at six locations in each of two years

Variety	*Location*					
	University	*Waseca*	*Morris*	*Crookston*	*Grand Rapids*	*Duluth*
1931						
Manchuria	1.349	2.441	1.371	1.995	1.647	1.447
Svansota	1.755	2.365	1.287	2.022	1.482	1.284
Velvet	1.993	2.510	1.306	2.065	1.151	1.314
Trebi	1.827	3.189	2.187	2.345	1.487	1.695
Peatland	1.637	2.426	1.492	2.078	1.734	1.599
1932						
Manchuria	1.344	1.672	1.717	1.647	1.106	1.127
Svansota	1.371	1.923	1.750	1.031	0.831	1.111
Velvet	1.339	1.868	1.940	1.602	1.610	1.122
Trebi	1.452	2.460	2.330	2.090	1.031	1.529
Peatland	1.402	1.800	2.158	1.261	1.337	1.567

9.4.1 Basic analysis

Yates and Cochran (1938) started with the simple variety means and basic analysis of variance of Table 9.2. Assuming that the V · L interaction is homogeneous, the standard error of the difference between the means of any two varieties is calculated as follows.

1. Express the expected values of the V · L, V · Y and units mean squares as linear functions of σ_{VL}^2, σ_{VY}^2 and σ^2.
2. Solve for the unknown components of variance.
3. If any estimate is negative set the corresponding component to zero, calculate a new pooled units sum of squares and return to stage 1.
4. Substitute the resulting estimates in a general formula for the standard error of a difference between two variety means.

The coefficients in the expressions for the three expected mean squares are:

Mean square	σ_{VL}^2	σ_{VY}^2	σ^2
V · L	n	0	1
V · Y		m	1
Units			1

where m is the number of locations and n the number of years. The general formula for the standard error is

$$\sqrt{\{2 \times (\sigma_{VL}^2/m + \sigma_{VY}^2/n + \sigma^2/mn)\}}$$

In the example, the V · Y mean square is 0.02023 with four degrees of freedom and the units mean square is 0.03861 with 20 degrees of freedom so we (i) set σ_{VY}^2 to zero and (ii) pool these two terms to give a new units mean square of $\sigma^2 = 0.03555$; the estimate of σ_{VL}^2 is given by $(0.06147 - 0.03555)/2 = 0.01296$ and the required standard error is 0.1012 (Table 9.2).

Table 9.2. Variety means, standard error of difference between two means (SED) and basic analysis of variance of Immer's data

Variety	*Mean (t/ha)*
Manchuria	1.572
Svansota	1.518
Velvet	1.652
Trebi	1.968
Peatland	1.708
SED	±0.1012

Effect	*df*	*Mean square*
L	5	1.1770
Y	1	1.0534
L · Y	5	0.3824
V	4	0.3682
V · L	20	0.06147
Units	24	0.03555
Total	59	

9.4.2 Modified analysis allowing for non-homogeneous V · L interaction

In this example, however, the V · L interaction is not homogeneous. Yates and Cochran (1938) noted that 'Trebi' yields varied more between locations than the yields of the other varieties. Accordingly, they divided the V · L degrees of freedom into two parts, one part representing the variation from location to location of the mean difference between 'Trebi' and other varieties and the other part representing the variation of differences between varieties other than 'Trebi'.

This is equivalent to defining a factor W (for groups of varieties) taking level 1 for 'Trebi' and level 2 for the other varieties, then subdividing the total V · L interaction into two parts: W · L interaction and residual V · L interaction. General formulae appropriate to two groups, one with s varieties, the other with t varieties are given below. The coefficients of components of variance in the expected values of mean squares are

Mean square	σ_{WL}^2	σ_{VL}^2	σ^2
W · L	$\lambda = 2nst/(s+t)$	n	1
V · L		n	1
Units			1

The standard error of a difference between two varieties in the same group is given by

$$\sqrt{\{2 \times (\sigma_{VL}^2/m + \sigma_{VY}^2/n + \sigma^2/mn)\}}$$

and for varieties in different groups is

$$\sqrt{\{2 \times (\sigma_{WL}^2/m + \sigma_{VL}^2/m + \sigma_{VY}^2/n + \sigma^2/mn)\}}$$

In this example, the residual V · L mean square is 0.02993, which is smaller than the units mean square of Table 9.2 so the residual V · L component is set to zero as well as the V · Y component. The W · L component is then estimated by

$$(0.1561 - 0.03339)/3.20 = 0.03835$$

The divisor in this expression is the value of λ when $n = 2, s = 1, t = 4$. Hence, the standard error of the difference between the means for any two varieties other than 'Trebi' is

$$\sqrt{(2 \times 0.03339/12)} = 0.07459$$

and the standard error of the difference between 'Trebi' and another variety is

$$\sqrt{\{2 \times (0.03835/6 + 0.03339/12)\}} = 0.1354$$

The second standard error is larger than the first because the difference between two varieties in different groups is more variable than the difference between two varieties in the same group. Table 9.3 gives a summary.

Table 9.3. Modified analysis of variance of Immer's data

Variety	*Mean (t/ha)*
Manchuria	1.572
Svansota	1.518
Velvet	1.652
Trebi	1.968
Peatland	1.708
SED	
pairs including Trebi	0.1354
other pairs	0.07459

Effect	*df*	*Mean square*
L	5	1.1770
Y	1	1.0534
L · Y	5	0.3824
V	4	0.3682
W · L	5	0.1561
Units	39	0.03339
Total	59	

9.4.3 Basic analysis using REML

Although simple in principle, calculations of this type are tedious in practice and not recommended for routine work on a large scale. Fortunately, the same results can be obtained more-or-less painlessly using REML.

We start with the basic analysis, for which the REML equivalent is to fit Equation 9.1 (REML 1 in Table 9.4). The estimated V · Y component is negative so this term is set to zero and the revised model fitted. The final estimated components of variance and standard errors are the same as those obtained in Section 9.4.1 using the analysis of variance. Fitting Equation 9.2 with random L, Y and L · Y does not affect the calculation of the standard errors for balanced data (REML 2 in Table 9.4).

Table 9.4. REML analysis of Immer's data

Effect	*REML 1* *V+L+Y+L · Y: V · L+V · Y*	*REML 2* *V:L+Y+L · Y +V · L+V · Y*	*REML 3* *V+L+Y+L · Y: W · L+V·L+V · Y*	*REML 4* *V:L+Y+L · Y+ W · L+V·L+V · Y*
Components of variance				
L	–	0.07687	–	0.1071
Y	–	0.02237	–	0.02237
L · Y	–	0.06937	–	0.06971
W · L	–	–	0.03835	0.03388
V · L	0.01296	0.01296	set to zero	set to zero
V · Y	set to zero	set to zero	set to zero	set to zero
Units	0.03555	0.03555	0.03339	0.03381
SED[a]				
Maximum	0.1012	0.1012	0.1354	0.1300
Minimum	0.1012	0.1012	0.07459	0.07506

[a] Variety means are given in Table 9.2 and Table 9.3.

9.4.4 Modified analysis using REML

REML can also be used to give the same results as the modified analysis of variance of Table 9.3 with factor W defining variety groups. The appropriate model is

$$V + L + Y + L \cdot Y : W \cdot L + V \cdot L + V \cdot Y \tag{9.3}$$

and the results are again in Table 9.4 (REML 3). As before the estimated V · Y component is negative. So, too, is the residual V · L component now that the W · L term has been included. Both components are set to zero.

Different results are obtained with random L, Y and L · Y, when the model is

$$V : L + Y + L \cdot Y + W \cdot L + V \cdot L + V \cdot Y \tag{9.4}$$

The differences arise because estimates of the W · L and units components of variance are slightly more accurate under Equation 9.4 than under Equation 9.3. Changes in standard errors are perceptible but of little practical importance.

9.5 METHOD OF DIRECT DIFFERENCES

9.5.1 Student's example

Student (1923) described a series of trials organized by the Irish Department of Agriculture to identify the best variety of barley to grow on Irish farms in the early years of this century. Plots were large, almost one hectare in size, and located on farmers' fields. The trials lasted six years, during which time seven varieties were tested. Two of these, 'Archer' and 'Goldthorpe', were grown every year. Others either were not chosen in the first year or were discarded before the series ended.

Student's paper provides yields of 'Archer' and 'Goldthorpe' for the 51 trials in which both were tested. Table 9.5 classifies these yields by the years and 20 combinations of farmer and place name. Student mentions that 18 farms were used in the trials but these are not clearly identified in the paper (some farms may have changed hands).

Table 9.5. Student's data: yields (t/ha) of two varieties of barley (*A* = 'Archer', *G* = 'Goldthorpe') at 20 locations in each of six years (converted from yields in Irish barrels per acre given in Student, 1923)

Location	*Variety*	*Year*					
		1901	*1902*	*1903*	*1904*	*1905*	*1906*
Ballinacurra	*A*	2.82	3.11	1.66	2.57	3.14	2.48
	G	1.76	2.98	1.82	2.98	3.28	2.43
Whitegate	*A*	2.56	3.52	2.20	2.68	2.94	2.65
	G	1.95	3.26	1.84	2.57	2.84	2.23
Bagnalstown	*A*	–	–	–	–	–	3.83
	G	–	–	–	–	–	3.48
Birr	*A*	–	3.11	2.46	2.81	3.69	2.75
	G	–	2.31	2.01	2.98	3.39	2.50
Monasterevan 1	*A*	–	–	–	2.62	–	–
	G	–	–	–	2.59	–	–
Monasterevan 2	*A*	–	–	–	–	3.64	2.42
	G	–	–	–	–	3.22	2.65
Monasterevan 3	*A*	–	–	–	–	–	3.14
	G	–	–	–	–	–	3.48
Nenagh	*A*	2.76	3.04	2.04	3.31	3.61	2.95
	G	2.51	3.36	2.12	2.89	3.92	2.21
Portarlington	*A*	–	–	–	3.03	3.03	–
	G	–	–	–	2.81	2.64	–
Thurles	*A*	3.80	–	–	–	–	–
	G	3.48	–	–	–	–	–
Tullamore	*A*	–	–	–	–	3.45	2.23
	G	–	–	–	–	2.67	2.18
Arnestown	*A*	–	–	1.33	–	–	–
	G	–	–	1.93	–	–	–
Castlebridge	*A*	–	2.81	3.12	2.29	2.86	2.92
	G	–	2.82	2.50	1.57	2.86	2.65
Enniscorthy	*A*	–	2.84	–	–	–	–
	G	–	2.98	–	–	–	–
New Ross 1	*A*	–	–	–	2.04	–	–
	G	–	–	–	1.76	–	–
New Ross 2	*A*	–	–	–	–	3.26	3.56
	G	–	–	–	–	3.42	3.09
Carlingford 1	*A*	–	–	2.95	–	–	–
	G	–	–	2.31	–	–	–
Carlingford 2	*A*	–	–	–	2.01	3.61	2.94
	G	–	–	–	2.37	2.82	3.20
Greenore	*A*	–	–	2.81	–	–	–
	G	–	–	1.96	–	–	–
Dunleer	*A*	–	–	–	3.03	3.62	3.61
	G	–	–	–	2.87	3.14	3.36

Student calculated the 51 differences between 'Archer' and 'Goldthorpe', averaged them and calculated their standard deviation and hence the standard error of the mean. The mean difference is 0.210 t/ha and its standard error is 0.0532 t/ha.

This calculation is valid only if the 51 differences are independent. A pair of differences may be correlated if they come from the same farm or from the same year. The point can be checked by fitting the REML model

$$V : L + Y + L \cdot Y + V \cdot L + V \cdot Y$$

to the data of Table 9.5. In spite of the large effects of years and the trials within years, the V · L and V · Y components of variance are zero (Table 9.6) so the 51 differences are virtually independent and Student's error calculation is correct.

Table 9.6. REML analysis of Student's data. Model V : L + Y + L · Y + V · L + V · Y

Variety	*Mean (t/ha)*	*Component of variance*	
A	2.882	L	0.0092
G	2.672	Y	0.1133
A – G (SED)	0.210 (±0.0532)	L · Y	0.1289
		V · L	set to zero
		V · Y	set to zero
		Units	0.0722

Student pointed out that the accuracy of his estimated mean difference depends on the correlation between the yields of the two varieties. The value of the correlation is 0.79. Borrowing terminology used in a slightly different context by Simmonds (1979, page 214) we call a correlation of this type a variety-on-variety or VV-correlation.

9.5.2 Drawbacks of direct differences

Student's analysis is an example of the method of direct differences. The essential feature of this method is that comparisons between two varieties are restricted to trials in which both occur. Only the differences are analysed. A simple variance calculation may suffice as in Student's example. This method has intuitive appeal and is frequently advocated. It has, however, serious drawbacks when applied to incomplete tables. In these cases, REML or the analysis of variance is needed to eliminate the effects on the differences of locations and years.

We demonstrate these drawbacks for dry matter yields of seven varieties of perennial ryegrass in five years of trials carried out by the National Institute of Agricultural Botany (NIAB), Cambridge, UK (Table 9.7). Each entry is a mean over seven trials. Only one variety, the control, was grown every year. Other varieties appeared in two or three years out of the five. Normally, the number of varieties is 30 or more including two controls and ten years' results are analysed.

Table 9.7. NIAB perennial ryegrass trials: mean yields over seven locations (t dry matter/ha) of seven varieties in five years

Variety	*Year*				
	1984	*1985*	*1986*	*1987*	*1988*
1 (control)	10.84	14.46	13.37	10.83	10.03
2	–	–	13.42	11.38	–
3	–	–	–	9.91	9.78
4	11.14	13.90	–	–	–
5	11.61	14.66	13.99	–	–
6	–	–	–	10.75	9.60
7	–	–	13.94	10.92	–

Table 9.8 gives the estimated mean direct differences d_{ik} between all pairs of varieties ($i, k = 1, 2 \ldots 7$ and $i \neq k$). For example, the mean yields of variety *1* and variety *5* over the three years in which both were grown are respectively 12.89 and 13.42 t/ha. Hence $d_{15} = -0.53$ or, equivalently, $d_{51} = +0.53$.

The main weaknesses of this method are inaccuracy and inconsistency (Patterson, 1978). Table 9.8 reveals the first of these. Direct differences are not available for six of the 21 pairs of varieties whilst others, such as d_{25}, are based on the yields in only a single year. The second weakness becomes apparent when constructing a list of variety means in rank order. The order of the top four varieties is clear with variety *5* first, then variety *7*, variety *2* and variety *1*, but all we know about the order of the bottom three varieties is that variety *6* takes precedence over variety *3* on the strength of superior performance in two years. Sometimes, although not in the present example, there are contradictions such as variety *A* is better than variety *B*, variety *B* is better than variety *C* and variety *C* is better than variety *A*.

Table 9.8. Mean direct differences d_{ik} (t dry matter per ha) between pairs of varieties of perennial ryegrass *i* and *k* (the table is skew symmetric so that $d_{ik} = -d_{ki}$)

Variety	*Variety*						
	1	*2*	*3*	*4*	*5*	*6*	*7*
1		−0.30	1.08	0.13	−0.53	0.26	−0.33
2	0.30		1.47	–	−0.57	0.63	−0.03
3	−1.08	−1.47		–	–	−0.83	−1.01
4	−0.13	–	–		−0.62	–	–
5	0.53	0.57	–	0.62		–	0.05
6	−0.26	−0.63	0.83	–	–		−0.17
7	0.33	0.03	1.01	–	−0.05	0.17	

9.6 INCOMPLETE TABLES AND THE ADJUSTMENT OF VARIETY MEANS

An alternative to the method of direct differences is to adjust the mean yields of varieties with missing entries. This can be done in several ways, of which we consider three: control varieties, FITCON and REML. We illustrate each method with reference to the ryegrass example of Section 9.5.2.

9.6.1 Control calculations

One version of the control method consists of calculating the mean direct differences between each variety and the control, as in the first column of Table 9.8, then adding to each of these 11.91, the control mean over all years. The results are in the third column of yields in Table 9.9.

Table 9.9. Adjusted mean yields (t dry matter/ha) of seven varieties of perennial ryegrass

Variety	*Method of adjustment*				
	None	*% of control*	*Difference versus control*	*FITCON V + Y:*	*REML V : Y*
1 (control)	11.91	11.91	11.91	11.91	11.91
2	12.40	12.23	12.21	12.13	12.13
3	9.34	10.66	10.82	10.83	10.81
4	12.52	11.84	11.78	11.79	11.81
5	13.42	12.43	12.44	12.43	12.44
6	10.18	11.61	11.65	11.66	11.64
7	12.43	12.21	12.24	12.16	12.16
Max. SED	–	–	0.369	0.343	0.342
Min. SED	–	–	0.213	0.206	0.205
Components of variance					
Y	–	–	–	–	3.279
Units	–	–	–	0.0682	0.0683

This method gives a single consistent set of variety means and avoids the worst inefficiencies of the method of direct differences. Nevertheless, inefficiency remains. The only varietal differences that use all of the available within-year information are those between pairs of varieties, such as 3 and 6, or 2 and 7, which were always tested together. Their standard error is 0.213. The least efficient comparisons, with the largest standard errors, are between varieties that were not grown in the same year.

Another version of the control method uses percentage relative yields instead of differences (Silvey, 1978). In the example, the average percentage yield of variety 2 is

$$100 \times (13.42/13.37 + 11.38/10.83)/2 = 102.7$$

which gives $(11.91 \times 102.7)/100 = 12.23$ t/ha for the adjusted mean (second column of yields in Table 9.9).

9.6.2 Basic FITCON and REML adjustments

The FITCON model V + Y uses information from both direct and indirect comparisons within years (cf. the analysis of incomplete block designs,

Section 4.7). The REML model V : Y uses information from comparisons between the years as well. The resulting variety means, standard errors and estimates of components of variance are in Table 9.9.

Both methods give the same mean for variety *1*, the only variety grown in every year. The means for variety *3* and variety *6* are adjusted upwards because these varieties were grown in the two poorest years. By contrast, yields of variety *4* and variety *5* are adjusted downwards.

The increased efficiency of FITCON and REML relative to the differences via controls method is reflected in the smaller standard errors assigned by these two methods.

The two components of variance estimated by REML provide useful information on efficiency, bearing in mind that the units variance is a mixture of V · Y variance and experimental error. An estimate of the average VV-correlation (Section 9.5.1) is given by the ratio $\sigma_Y^2/(\sigma_Y^2 + \sigma^2)$, which in this case is 0.98, an unusually high value.

REML adjustments are smaller than FITCON adjustments. We describe this phenomenon by saying that REML moderates the FITCON adjustments. The degree of moderation depends on the VV-correlation; when the correlation is large, as in the present example, very little moderation is needed.

9.6.3 FITCON and REML analysis allowing for non-homogeneous interaction

As noted in Section 9.4.2, non-homogeneity of interaction in an incomplete varieties × locations table affects not only the estimation of standard errors but also the estimation of variety means. Clearly, this is a serious problem. Fortunately, it also appears to be rare in UK variety testing (Silvey, 1978).

The discussion is illustrated using the yields of seven varieties of potato at 16 NIAB locations in 1975 (Table 9.10). They were obtained by omitting the yields of two varieties, 'Cara' and 'Desiree', at eight locations from an originally complete table (Patterson, 1982).

Table 9.10. NIAB potato trials: yields (t/ha) of seven varieties at 16 locations, 1975

Variety	*Location*															
	1	*2*	*3*	*4*	*5*	*6*	*7*	*8*	*9*	*10*	*11*	*12*	*13*	*14*	*15*	*16*
Cara	–	–	–	–	–	–	–	–	35	53	44	61	49	59	59	31
Desiree	–	–	–	–	–	–	–	–	34	79	54	56	41	28	40	34
Drayton	39	43	31	40	31	41	56	34	31	69	32	54	36	30	44	36
King Edward	37	47	29	43	26	47	51	41	33	82	30	54	32	29	42	38
Estima	46	46	29	45	29	46	55	32	28	74	50	62	39	25	45	31
Majestic	40	49	39	46	31	49	54	40	33	72	38	61	34	29	39	37
Pentland Crown	44	49	30	44	47	49	59	49	37	88	50	62	53	38	45	40

The unadjusted means for the seven varieties are in the first column of figures in Table 9.11. The second column gives the FITCON means; 'Cara' and 'Desiree' means are adjusted downwards because the trials including these two varieties yielded more than average. Otherwise the means are unchanged. The third column gives the variety means, components of variance and standard errors of typical varietal differences when the model V : L is fitted; we call this model REML 1. The average VV-correlation is 0.76, much smaller than in the ryegrass example. Even so, REML 1 has moderated the FITCON adjustments by only 6%.

Table 9.11. Adjusted mean yields (t/ha) of seven varieties of potato

Variety	*Model*				
	Unadjusted	*FITCON* $V + L$	*REML 1* $V : L$	*REML 2* $V : L + W \cdot L$	*Omitting 'Cara'* $V : L$
Cara	49.76	47.50	47.58	49.37	–
Desiree	45.64	44.38	44.45	44.42	44.41
Drayton	40.30	40.30	40.30	40.30	40.30
King Edward	41.31	41.31	41.31	41.31	41.31
Estima	42.42	42.42	42.42	42.42	42.42
Majestic	42.93	42.93	42.93	42.93	42.93
Pentland Crown	49.70	49.70	49.70	49.70	49.70
Effect	Components of variance				
L	–	–	119.7	41.89	143.6
W · L	–	–	–	89.51	–
Units	–	37.58	37.54	20.29	20.35
Variety difference	Standard errors of differences between means				
Cara versus Desiree	–	3.06	3.06	4.85	–
Cara versus Drayton	–	2.74	2.74	4.71	–
Desiree versus Drayton	–	2.74	2.74	2.01	2.02
Drayton versus Estima	–	2.17	2.17	1.59	1.60

FITCON and REML 1 are valid only when the V · L interaction is homogeneous. Following Yates and Cochran's (1938) example, we inspect Table 9.10 and notice that much of the interaction might be explained by the unusual performance of 'Cara': it was poor in relation to the other varieties at location 10, good at locations 14 and 15 and average in five other trials. This suggests that we should fit a REML model with two groups of varieties, one consisting of 'Cara' alone and the other of the remaining six varieties; the model is V : L + W · L, which we call REML 2.

The REML 2 components of variance in Table 9.11 show that W · L accounts for a large part of the V · L interaction (which is included in the units variance). This is reflected in the large standard errors of differences involving 'Cara' and the small standard errors of other differences. Interpretation of the components in terms of VV-correlations is a little more difficult than in

REML 1. The ratio $(\sigma_L^2 + \sigma_{WL}^2)/(\sigma_L^2 + \sigma_{WL}^2 + \sigma^2)$, estimated by 0.86, is the average VV-correlation within groups, i.e. between varieties other than 'Cara'. The average correlation between 'Cara' and other varieties is $\sigma_L^2/(\sigma_L^2 + \sigma_{WL}^2 + \sigma^2)$; the estimate is only 0.29.

Omitting 'Cara' from the analysis and fitting V : L gives almost the same results as REML 2 for the remaining varieties. The VV-correlation is now 0.89.

A possible explanation for the anomalous behaviour of 'Cara' is that the 1975 summer months were unusually dry and sunny; 'Cara' is a late-maturing variety and was not affected by the drought in the same way as other varieties.

9.6.4 Estimation of varietal sensitivity to changes of environment

Yates and Cochran (1938) suggest that we can often detect heterogeneity of V · L interaction in a complete varieties × locations table by plotting the yields for each variety against the location means. When the table is incomplete the yields can be plotted against location effects given by the model V + L. The slope of each regression line estimates the sensitivity of a variety to differences between locations.

A more refined but closely related method is to fit the non-linear model:

$$y_{ij} = \alpha_i + \beta_i\,\phi_j + e_{ij} \tag{9.5}$$

where y_{ij} is the yield of variety i at location j, α_i is the mean for variety i, ϕ_j is the effect of location j, β_i is the sensitivity of variety i, and e_{ij} is $N(0,\sigma^2)$, a normally distributed error variable with mean 0 and variance σ^2. It is convenient to scale the β so that their average over all varieties is 1. We can then refer to varieties as having average ($\beta_i \approx 1$), less-than-average ($\beta_i < 1$) or more-than-average ($\beta_i > 1$) sensitivity. Sensitivity is a relative rather than an absolute measure of variability. It depends on the other varieties included in the analysis, on the particular set of conditions sampled and even on the way the trials are managed. Digby (1979) describes an iterative method for fitting Equation 9.5 and Williams and Matheson (1994) provide a GENSTAT algorithm.

In the present example, the sensitivity of 'Cara' is significantly smaller than average (Table 9.12). The units variance from Equation 9.5 is smaller than the units variances from FITCON and REML 1 but appreciably larger than the REML 2 units variance in Table 9.11. This suggests that the differences between the sensitivities do not account for all of the W · L interaction and is a warning against using Equation 9.5 to estimate variety means. Omitting the yields of 'Cara' and refitting the model gives no apparent differences between the sensitivities of the remaining varieties; the error variance is now almost identical to the REML 2 units variance. This suggests that residual interaction in REML 2 is reasonably homogeneous.

In my view sensitivities should be calculated as a matter of routine to identify varietal heterogeneity in V · L interaction. Useful additional checks are provided by actual VV-correlations, insofar as these can be calculated individually, and standard deviations over locations. REML itself can be used directly

Table 9.12. Sensitivities of seven varieties of potato

Variety	*All varieties*		*Omitting 'Cara'*	
	Sensitivity	*Standard error*	*Sensitivity*	*Standard error*
Cara	0.382	0.133	–	–
Desiree	1.130	0.133	1.014	0.098
Drayton	0.968	0.116	0.873	0.087
King Edward	1.198	0.116	1.094	0.087
Estima	1.194	0.116	1.076	0.087
Majestic	1.006	0.116	0.919	0.087
Pentland Crown	1.123	0.116	1.023	0.087
Units variance	30.46		20.36	

to check for heterogeneity due, for example, to differences between control and other varieties, varieties of different age, early and late-maturing varieties, varieties differing in ploidy and so on.

9.6.5 Sensitivity analysis with random location effects

Digby's method divides total V · L variation into two parts, one due to differences in sensitivity and the other to residual V · L interaction. Only the second part is treated as random. In consequence, standard errors assigned to differences between the predicted variety means are too small unless the varieties are equally sensitive. This difficulty can be overcome by fitting model B from Patterson and Silvey (1980, Section 5.2.2):

$$y_{ij} = \alpha_i + \beta_i \phi_j + e_{ij} \tag{9.6}$$

where e_{ij} is $N(0,\sigma^2)$ and ϕ_j is $N(0,\sigma_c^2)$. This is equivalent to Equation 9.5 except that the location effects ϕ_j are randomly and Normally distributed. Thus, we have to fit a non-linear model with two error terms; this requires an extension of the existing REML algorithm.

9.7 AN INCOMPLETE THREE-WAY TABLE

The first example (Section 9.4) analysed a summary table with no missing values and structure V × L × Y. For contrast, we now consider an incomplete table with the locations changing from year to year and structure V × (Y / L).

9.7.1 Data

Table 9.13 gives the yields of the 29 varieties of sugar beet considered by NIAB for recommendation to farmers in 1992. For full recommendation, a variety would have to perform well over the five years 1987–91; varieties completing their third or fourth years in 1991 were considered for provisional recommendation.

Of the varieties in Table 9.13, varieties *1* to *4* were already on the list of fully recommended varieties, variety *5* had been provisionally recommended since 1990 and variety *6* since 1991. The number of trials in each year varied from 11 to 16. Second-year and older varieties were included in all trials but first-year varieties were restricted to a subset of seven trials.

Routine analysis consisted of arranging the data for the five years in a two-way varieties × trials table and fitting constants for varieties and trials. This method is commonly referred to as single-stage FITCON. It has been discussed by Silvey (1978) and Patterson (1978).

9.7.2 Basic analysis

The main purpose of the example is to demonstrate the efficient REML basic analysis. The formula for the appropriate model for a V × (Y / L) table is V : Y + Y · L + V · Y (Section 9.3). In the absence of an L main effect, Y · L represents the effects of the locations within the years; it is not an interaction.

This method will be compared with single-stage FITCON, which is equivalent to fitting the model V + Y · L. A third method will also be considered: two-stage FITCON (Patterson, 1978; Patterson and Silvey, 1980). Two-stage FITCON consists of the following.

1. Calculate the variety means for each year and adjust for the differences between the locations when necessary. Thus, variety means for 1987 are obtained using the FITCON model V + L; similarly for 1988 and 1989. For 1990 and 1991 the data are complete and ordinary variety means suffice.
2. Assemble the results in Table 9.14 and apply FITCON again, this time using V + Y as the model.

Estimates of the variety means given by the three methods are in Table 9.15, together with maximum and minimum standard errors for the differences between two varieties. The two-stage FITCON means have been centred so that all three sets of figures have the same mean over all varieties.

The estimated Y component of variance in the basic REML model is negative so this model is replaced by V : Y · L + V · Y. Maximum discrepancies between the means given by the two approximate methods and the REML means are 0.03 in single-stage FITCON and 0.01 in two-stage FITCON. The two-stage method gives reasonable approximations for the standard errors but single-stage FITCON estimates of error are gross underestimates.

More generally, when the numbers of trials are the same in each year there is little to choose between single-stage FITCON and two-stage FITCON in terms of efficiency; one gives equal weight to each trial, the other equal weight to each year. However, when the numbers vary greatly, two-stage FITCON can be very inefficient (Patterson, 1978). Single-stage FITCON

Table 9.13. NIAB sugar beet trials: yields (t sugar/ha) of nine varieties at up to 16 locations in five years

Variety	*Location*															
	1	*2*	*3*	*4*	*5*	*6*	*7*	*8*	*9*	*10*	*11*	*12*	*13*	*14*	*15*	*16*
Year 1987																
1	11.75	9.91	11.57	9.31	10.69	9.59	9.67	11.33	9.87	11.78	10.35	10.78	9.92	9.06	10.22	9.04
2	11.68	9.45	11.66	9.43	11.76	9.47	9.32	11.50	9.28	11.94	10.46	10.92	9.47	9.88	10.23	9.37
3	11.77	9.50	12.23	9.56	10.69	9.99	9.50	11.69	9.34	11.73	10.30	11.00	9.55	9.84	10.13	9.44
4	11.55	9.52	12.18	9.35	10.34	9.49	9.43	11.36	9.68	11.11	10.53	9.86	9.52	9.13	10.23	9.66
5	11.86	9.70	12.18	9.90	11.10	9.04	9.20	–	–	–	–	–	–	–	–	–
6 to 9	–	–	–	–	–	–	–	–	–	–	–	–	–	–	–	–
Year 1988																
1	11.48	11.71	11.25	11	9.2	9.73	9.28	10.56	12.06	11.9	11.18	10.26	10	10.63	11.32	10.59
2	10.85	11.77	11.02	10.4	9.18	10.06	7.35	10.37	12.23	12.34	11.41	10.44	9.64	10.03	11.16	10.13
3	11.59	12.24	11.73	12.27	9.6	9.2	9.02	10.93	11.87	11.6	10.86	10.39	9.76	10.77	11.01	10.72
4	11.2	12.26	11.29	10.7	9.75	9.92	7.73	11.05	11.95	12.42	11.54	10.71	9.33	10.76	11.25	10.31
5	11.76	12.11	11.37	12.29	9.94	10.03	7.98	10.53	11.99	11.8	11.9	10.56	9.74	10.47	11.28	10.07
6	10.8	11.99	12.12	12.2	10.29	9.51	7.85	–	–	–	–	–	–	–	–	–
7 to 9	–	–	–	–	–	–	–	–	–	–	–	–	–	–	–	–
Year 1989																
1	9.26	7.29	9.46	12.39	11.71	11.94	6.91	9.40	9.02	10.33	15.89	9.39	9.19	12.02	9.19	11.32
2	9.16	7.34	9.82	12.14	12.37	11.13	6.65	9.38	9.43	10.85	15.03	9.03	9.08	12.28	9.55	11.43
3	9.35	7.18	9.80	12.14	11.74	11.46	6.79	9.72	9.26	10.81	16.18	9.52	9.62	11.89	9.52	11.60
4	9.18	7.88	9.96	13.14	12.76	12.15	7.14	11.04	9.49	10.18	14.93	10.21	9.44	12.25	10.32	12.32
5	9.56	9.20	10.37	12.76	13.07	12.94	7.04	9.69	9.36	10.44	14.86	9.52	9.92	12.83	10.42	12.54
6	9.99	7.21	10.15	11.99	12.55	11.73	6.28	9.85	9.20	9.88	14.66	9.96	9.13	12.01	9.60	10.89
7	9.93	6.70	10.87	13.02	12.86	10.76	7.23	–	–	–	–	–	–	–	–	–
8	9.23	7.49	9.79	11.60	12.69	11.31	7.07	–	–	–	–	–	–	–	–	–
9	9.97	7.33	10.70	12.42	11.79	11.51	7.51	–	–	–	–	–	–	–	–	–

Year 1990													
1	7.44	9.14	12.88	11.29	9.63	9.26	15.00	9.68	9.01	9.90	9.28	–	–
2	7.56	9.07	13.58	11.20	9.07	9.52	13.66	9.78	7.25	9.45	9.59	–	–
3	7.68	9.21	12.91	11.60	9.33	9.33	13.53	9.62	9.50	9.67	9.91	–	–
4	9.06	9.77	14.10	11.00	9.27	9.64	14.60	9.80	9.04	9.64	9.52	–	–
5	7.95	9.07	15.30	12.12	9.69	9.92	15.12	9.41	9.57	9.92	9.88	–	–
6	7.75	9.79	15.27	11.36	9.24	9.62	15.01	9.36	9.18	9.79	9.47	–	–
7	9.38	9.81	12.76	11.60	7.69	9.79	15.19	9.58	9.05	7.95	9.07	–	–
8	7.25	9.99	13.37	10.35	9.34	7.86	14.17	7.52	9.26	9.26	7.46	–	–
9	7.42	9.86	13.41	11.14	9.58	9.23	14.97	9.32	9.51	9.73	9.91	–	–
Year 1991													
1	10.73	10.14	7.68	10.57	10.21	9.94	12.71	10.32	9.17	11.75	9.97	10.82	13.50
2	10.66	11.54	9.22	10.90	10.58	10.35	12.29	11.05	9.86	11.91	10.15	10.98	14.38
3	10.38	10.34	7.86	10.98	10.52	9.04	11.42	10.75	9.19	11.09	10.01	10.88	13.53
4	10.85	10.82	9.76	11.74	10.72	10.98	12.59	11.35	10.14	11.75	9.95	11.23	13.90
5	10.89	10.72	7.73	11.21	9.93	10.20	12.26	10.78	9.48	11.19	10.16	10.73	13.55
6	10.84	10.67	7.66	12.27	10.47	10.80	11.68	11.67	9.87	12.07	10.41	11.30	14.92
7	11.31	11.40	9.46	11.82	10.92	11.40	13.19	11.12	9.68	12.14	10.68	11.04	13.50
8	10.24	11.07	9.38	10.04	10.03	10.08	12.48	10.72	9.92	11.25	9.86	10.76	13.55
9	11.13	10.62	9.50	11.60	11.23	10.44	12.72	11.47	9.22	11.79	10.16	10.95	13.99

Table 9.14. FITCON means (t sugar/ha), 1987–91

Variety	*Year*				
	1987	*1988*	*1989*	*1990*	*1991*
1	10.115	10.697	10.294	9.774	10.578
2	10.239	10.524	10.167	9.521	10.990
3	10.266	10.785	10.411	9.754	10.461
4	9.966	10.761	10.774	9.949	11.137
5	10.281	10.864	10.782	10.359	10.679
6	–	10.941	10.192	9.894	11.125
7	–	–	10.464	9.716	11.282
8	–	–	10.294	9.348	10.568
9	–	–	10.444	10.098	11.063
Max. SED	0.1432	0.1732	0.2344	0.2141	0.1413
Min. SED	0.1066	0.1301	0.1551	0.2141	0.1413

Table 9.15. Predicted variety means (t sugar/ha) of nine varieties of sugar beet

Variety	*Single-stage FITCON*	*Two-stage FITCON*[a]	*Basic REML*
1	10.32	10.31	10.31
2	10.31	10.31	10.31
3	10.37	10.36	10.36
4	10.54	10.54	10.54
5	10.62	10.61	10.62
6	10.47	10.50	10.50
7	10.58	10.55	10.55
8	10.11	10.13	10.12
9	10.62	10.60	10.60
Max. SED	0.112	0.180	0.188
Min. SED	0.074	0.139	0.138
Effect	Components of variance		
Y	–	–	set to zero
Y · L	–	–	2.633
V · Y	–	–	0.036
Units	–	–	0.166

[a] Centred by adding 0.02 to each mean.

almost always underestimates error; the example using Student's data provides one of the few exceptions, possibly because experimental errors were large by present-day standards. Two-stage FITCON underestimates errors when the same locations are used each year.

A modified weighted analysis for the second stage of the two-stage method was described by Patterson and Nabugoomu (1992) who, however, considered that time spent refining a basically inefficient method was time wasted.

9.7.3 Further analysis

Whichever method we use to combine the results over the years, it is essential to look at each year separately. In particular, sensitivities to differences between locations, VV-correlations, standard deviations and residuals should be calculated routinely.

These calculations reveal little of note apart from a significantly low sensitivity for variety *3* in 1990. This appears to be an isolated phenomenon with no discernible effect on over-year means. There is no other evidence of heterogeneity of V · L error.

V · Y error can be checked similarly but less usefully, because the number of years is small. External information on varieties and years may, however, suggest possible sources of heterogeneity. For example, the years fall into two groups: 1987–88 and 1989–91. Rainfall was average during the first two years but was exceptionally low in the last three. We may reasonably enquire whether the dry weather affected some varieties more than others.

This question can be investigated by (i) adding a random term for varieties × groups of years to the basic REML model and refitting, (ii) identifying the long-term relative frequencies of wet and dry years and (iii) calculating the predicted variety means for a population of years with these frequencies. The varieties × groups of years component of variance is small; variety means have slightly increased standard errors but are otherwise unaffected.

9.7.4 Changes in variety performance

It is important to check that there are no trends or other systematic changes in the yields of varieties. A sugar beet variety might change as a result of, for example, cross pollination. For this reason, NIAB keeps varieties of this (and some other crops) in trial even after they have been fully recommended (Silvey, 1978, page 388). This raises the question: how should the checks be made?

One possibility is to include an appropriate age factor in the calculations. Armed with information provided by the breeders we define a seed class factor S with levels *b*, *p* and *c* corresponding to the use of breeders' seed during the first three years of testing, pre-commercial seed in the fourth year and commercial seed in the later years (Table 9.16). Then we fit the model:

$$\mathrm{V + S \cdot V : Y + Y \cdot L + V \cdot Y} \qquad (9.7)$$

This allows us to estimate the yield change associated with each change of seed. The residual V · Y component of variance is 0.023, compared with 0.036 on the basic REML model. Variety means from each of the three seed classes are in Table 9.17. Predictions are not available for the yields of varieties *6*, *7*, *8* and *9* when grown from commercial seed.

The evidence of the example is inconclusive. The estimated seed effects in Table 9.17 are large enough to be of agricultural importance, but too small to be of statistical significance at conventional levels of probability; they may also be affected by unknown confounding factors which happen to be correlated with age.

Table 9.16. Age in 1991 of nine varieties of sugar beet and their seed classes in 1987–91

Variety	Age in 1991	Seed class[a] in				
		1987	1988	1989	1990	1991
1	13	*c*	*c*	*c*	*c*	*c*
2	9	*c*	*c*	*c*	*c*	*c*
3	7	*b*	*p*	*c*	*c*	*c*
4	6	*b*	*b*	*p*	*c*	*c*
5	5	*b*	*b*	*b*	*p*	*c*
6	4	–	*b*	*b*	*b*	*p*
7	3	–	–	*b*	*b*	*b*
8	3	–	–	*b*	*b*	*b*
9	3	–	–	*b*	*b*	*b*

[a] *b*, breeders' seed; *p*, pre-commercial seed; *c*, commercial seed.

Table 9.17. Predicted variety means (t sugar/ha) from three seed classes using Equation 9.7

Variety	Seed class (see Table 9.16)			
	b	*p*	*c*	*c*−*b*
1	–	–	10.32	–
2	–	–	10.31	–
3	10.48	10.46	10.29	−0.19 ±0.245
4	10.32	10.81	10.65	0.33 ±0.215
5	10.63	11.01	10.24	−0.39 ±0.242
6	10.43	10.68	–	–
7	10.57	–	–	–
8	10.14	–	–	–
9	10.62	–	–	–

The statistician can do little more than recommend further investigation. Analysis of larger sets of data with more varieties and more years would help but ultimately the only way to determine seed effects with accuracy and assurance is to compare old and new seed in the same trials.

9.8 OTHER MODELS

The key to the successful use of the REML algorithm in analysing series of variety trials lies in the choice of model. Basic models are determined by the structure of the data and knowledge of variation in earlier trials (Talbot, 1984). Section 9.3 gave examples of the recommended basic models.

When two series of trials are analysed together, a mixture of models may be appropriate. Suppose, for example, that one series uses the same locations every year whilst in the other they are chosen independently each year. The structure V × L × Y is then appropriate to one series, the structure V × (Y / L) to the other. In some circumstances, we have to modify the recommended basic model. If, for example, the current estimate of a component of variance is zero or negative, the corresponding error term must be dropped. This action may also be appropriate when the component is small both in the current series and in earlier series. Relying solely on tests of significance is not recommended. This practice leads in the long run to an underestimation of error.

A basic model has a simple error structure with the same set of VV-correlations (over locations, over years and residual) for every pair of varieties. This is adequate for many series, possibly because experimenters recognize the impracticability of testing very different varieties in the same small-plot trials. Sometimes, however, we need other models, such as those used in Sections 9.4.4 and 9.6.4, to allow for two levels of variation: between varieties in the same group and between varieties in different groups. Special analysis is also required when locations are stratified by regions.

Models based on methods described by Hemmerle and Downs (1978) extend the number of possibilities. These models include one or more supplementary environment factors that use the same levels as the original environment factor (e.g. location or years) but are defined only for the most variable varieties.

For example, we know that 'Cara' is less variable than other varieties in the potato data of Table 9.10 so we define a second locations factor, M say, for varieties other than 'Cara'. Fitting the model V : L + M gives roughly the same variety means as REML 2 in Table 9.11 but a narrower range of standard errors of differences (1.61 to 3.92 instead of 1.59 to 4.85); convergence is slow. Patterson and Nabugoomu (1992) give another example.

Many other models and methods of analysis have been suggested over the years; Freeman (1973) reviews the earliest of these. Gauch (1988) describes recent developments and in particular the AMMI method (see Section 8.4.1). Although outwardly similar, AMMI and Equation 9.6 have different objectives. The main purpose of AMMI is to summarize the responses of the varieties to a particular set of conditions. Equation 9.6 is solely concerned with prediction and to this end treats all varieties × environments interactions as sampling error.

ACKNOWLEDGEMENT

I am grateful to the National Institute of Agricultural Botany for allowing me to use examples of their data to illustrate points of statistical technique.

10
Resource allocation for selection systems

M. TALBOT

10.1 INTRODUCTION

Throughout the world a very large amount of money is devoted to plant variety evaluation. One of the larger US maize breeding companies sows 700 000 small plots spread across 200 locations annually. If we assume a plot has a variable cost of US$10, this represents an operating cost of US$7 million annually. In the UK, national crop advisory agencies sow more than 20 000 large plots every year, a plot having a variable cost of approximately US$50, making a total annual testing bill of some US$1 million.

Since only a tiny proportion of all plots are allocated to genotypes which are eventually successful in agriculture, there must be considerable savings from minimizing effort on less useful material. Such savings can come from discriminating good from mediocre genotypes as early as possible in the selection process. To design screening procedures to achieve this we need to know the relative contribution of the components of GE variation and we also require a criterion to measure progress in selection. This chapter explains how to apply estimates of GE variation and measures of efficiency to produce selection systems which help to concentrate effort on the better genotypes. The focus is yield, but the principles can be applied to most characters, and also to indices which combine more than one character.

10.2 VARIATION IN TRIALS SYSTEM

10.2.1 Components of variance

Although analysis of variance is a widely used tool for exploring biological data, few standard statistical packages routinely partition the analysis of variance mean square terms into components of variance. For example, in a randomized-block field experiment, the blocks mean square represents the

Statistical Methods for Plant Variety Evaluation. Edited by R.A. Kempton and P.N. Fox. Published in 1997 by Chapman & Hall, London. ISBN 0 412 54750 3.

sum of two variance components – one due to the effects of plot-to-plot variation and the other due to the effects of additional block-to-block variation. Knowledge of the relative size of the two variance components is useful in understanding experimental variation, and also in efficient allocation of experimental resources for the future. In complex data sets, such as those arising from series of trials, separating the total variance of all observations into parts associated with each of the effects aids understanding the importance of each factor.

There are several reasons why components of variation are not routinely produced, one of which is the computational problem of dealing with incomplete data sets. Since genotypes are continually entering and leaving trials, the genotypes × locations × years data matrix is incomplete. Estimating components of variance from such tables in the past has been difficult and the practice has been to avoid the problem by restricting the analyses to complete sets of data where standard analysis of variance techniques are available. These involve computing mean squares for the main effects and interactions in the analysis of variance, equating the mean squares to their expectations, and solving for the unknown variance components. While the analysis of variance method provides unbiased estimates, it does not make the best use of available information when the data matrix has empty cells. Advances in computing have enabled more efficient maximum likelihood methods such as the residual maximum likelihood (REML) approach proposed by Patterson and Thompson (1974) and outlined in Chapter 9.

10.2.2 Variation within trials

A starting point for improving the effectiveness of trials is reducing within-trial measurement error by design and analysis (see Chapters 4–7). For UK cereal trials it has been estimated that replacing complete block with alpha designs has produced gains in within-trial efficiency equivalent to that of an extra replicate, and spatial methods have been shown to produce similar or greater gains. With alpha designs, plant breeders can usually estimate yield differences between two genotypes in trials with a standard error of approximately 5%. This standard error represents a measure of the similarity in the relative performance of the genotypes if a trial is repeated at that location under the same conditions of management, weather, pests and disease. While it is important to ensure that within-trial error is controlled, particularly for studying GE interactions, nevertheless a single trial is of limited value in predicting the average performance of genotypes when grown more widely. For example, the results from a typical trial in northern European conditions estimate how two genotypes will yield throughout a country with a standard error of difference of approximately 12%.

In spite of substantial between-trial variation, good genotypes can be identified by combining information from several environments. Breeders will sometimes discard an unpromising line on the results from a single environment but only take important decisions on the promotion of better genotypes with results from trials replicated over locations and years.

10.2.3 Variation between trials

The major environmental factors affecting yield performance in a series of trials can be classified under the heading of location and year. The term 'location' embraces all of those effects which contribute to making the performance at one place different from that at other places each year. These effects are primarily associated with soil and agronomic management, but some climatic effects may also be involved. Year effects are principally caused by differences in weather but may also be due to variation in disease levels and to changes across the seasons in husbandry practices which apply at all locations.

Year and location factors do not act independently. A prolonged period without rain will have different consequences for trials on sandy soils than for those on heavier land. The positioning of a trial within a location may change from year-to-year and local husbandry practices may also change. All of these effects can be represented by a location × year interaction term.

Of most interest here is how the relative performance of genotypes changes among environments. One environmental factor will not influence relative performance to the same extent as another. Hence separate genotype × location (G · L), genotype × year (G · Y) and genotype × year × location (G · Y · L) terms are required. The G · L component results from the relative performance of genotypes changing from location-to-location in a way that is similar each year. The G · Y variance arises from differences in genotype performance between seasons which are apparent at all locations. The G · Y · L term represents genotype differences which change from location-to-location to an extent that is dependent on the season, or genotype differences which are affected by seasonal changes at some locations more than others.

Interactions as described above are based on a model background which regards effects as additive; that is the difference between two genotypes are approximately the same whether average yields are 2 or 8 t/ha. In practice this is rarely the case and the background model will tend to lie somewhere between additive and multiplicative. However, the additive model provides a reasonable working approximation in most situations, unless the range of trial yields exhibit four-fold or greater differences. Then one might want to consider, for example, a logarithmic transformation of the data (see Chapter 4), or introduce terms into the model to represent the regression of genotype yields on environment means (see Chapter 8). In their study of GE interactions in barley, Finlay and Wilkinson (1963) combined both approaches.

The contribution of the GE components will vary from crop to crop and will depend also on the environments in which the trials are sown. Nevertheless, some generalizations are possible. It is common experience that year effects, and consequently G · Y interactions, are larger than the location effects and G · L interactions. It is also true in general that G · Y · L interactions are larger than those for G · Y or G · L. This pattern is illustrated in GE components of variance from national barley and wheat variety testing programmes in the UK, Germany and Spain (Table 10.1). The components are presented as standard deviations (square roots of the variance) and expressed as a percentage of the mean yield. Despite the different conditions in these countries GE components are of similar size.

Table 10.1. Components of GE variation (% CV) for barley and wheat in official variety performance trials from three European countries

Component	*UK*	*Germany*	*Spain*	*Average*
G·Y	3.1	2.5	4.2	3.3
G·L	2.2	2.4	2.8	2.5
G·Y·L	5.6	5.2	5.5	5.4
Experimental error	5.4	6.8	10.1	7.4
No. of years	5–12	8	4–5	
No. of locations/year	20–26	15–17	20–27	

What happens when we broaden the range of environments to cover several countries? A Joint European Spring Barley Trial (JESBT) (Riggs, 1986) involved 30 varieties of spring barley grown in 13 breeder's trials covering nine European countries in each of three years. The components of variance are compared (Table 10.2) with results from UK official spring barley trials for genotypes which have been more highly selected for local conditions.

Table 10.2. Components of GE variation (% CV) in European (JESBT) and UK spring barley trials

Component	*JESBT*	*UK*
G·Y	0.8	3.3
G·L	3.2	2.3
G·Y·L	4.7	4.9

In the JESBT, differences from year-to-year in genotype performance that are apparent at all locations (G · Y) tend to be small. Also, as one might expect from the greater geographic range sampled, factors associated with locations have a more important role in the JESBT than in the UK series. However, despite the wider range of environments covered by the JESBT, the G · Y · L term is very similar in the two series. Table 10.2 suggests that broadening the range of environments may dilute the effect of the years but enhance somewhat the contribution of the locations. At the same time, the combination of the years and locations within a country, if sufficiently widely sampled, can often provide as extreme genotype responses as a geographically more dispersed set of trials.

In the following section we concentrate on the allocation of resources within a country and will use the average results in Table 10.1 for illustration.

10.3 PRECISION OF GENOTYPE MEANS

Components of variance can be used to calculate the precision of genotype estimates based on means from a series of trials. To do this we use a fundamental statistical formula for the variance of two independent measurements, x and y:

$$\mathrm{VAR}(x + y) = \mathrm{VAR}(x) + \mathrm{VAR}(y) \quad (10.1)$$

i.e. the variation of the sum of two independent measures is equal to the sum of their individual variations.

The mean difference in yield over a series of trials between a new genotype and one or more checks provides an estimate of the value that the new genotype will have if grown over the range of conditions sampled by the trials. This estimate is subject to uncertainty because future environmental conditions cannot be predicted exactly. Suppose that the new genotype is tested at l locations in y years and that the same c checks are used throughout. All genotypes, both new and check, are grown in r plots at each location. The standard error of the mean difference between a new genotype and checks is

$$\mathrm{SED} = \sqrt{[(1 + c)\,\mathrm{VAR}/c]} \quad (10.2)$$

where

$$\mathrm{VAR} = \frac{\sigma_{\mathrm{GY}}^2}{y} + \frac{\sigma_{\mathrm{GL}}^2}{l} + \frac{\sigma_{\mathrm{GYL}}^2}{ly} + \frac{\sigma^2}{lyr} \quad (10.3)$$

In Equation 10.3, σ_{GY}^2, σ_{GL}^2 and σ_{GYL}^2 are the GE components of variance described above and σ^2 is the within-trial, between-replicate variance. While the GE terms tell something about how genotypes respond to changes in farming conditions, the σ^2 term is primarily a measure of the repeatability of observations within a trial.

Equation 10.3 illustrates another important statistical principle, namely, that if the errors of the components of a measure are additive then the precision of the measure as a whole can best be improved by reducing the largest component of variance. For example, consider a trial system for cereals based on averages of the variance components in Table 10.1, and with ten locations, two years and three replicates per trial. The contribution to VAR of the four components of variation in Equation 10.3 are $(3.3)^2/2$, $(2.5)^2/10$, $(5.4)^2/20$, $(5.0)^2/60$, where we assume an average within-trial coefficient of variation of 5%. Hence the total variance per genotype is

$$5.44 + 0.62 + 1.46 + 0.42 = 7.94$$

This example shows why the G · Y variance is important in practice. More than two-thirds of the total variance is due to this component. No matter how many locations are used, the total variance cannot be reduced below 5.44 without increasing the number of years.

Using Equation 10.3 we can derive critical differences. A critical difference is the difference between one genotype and another which, if the true difference is zero, will be exceeded by chance in trials in a proportion, α, of cases. The critical difference, when expressed as a percentage of mean yield, is calculated as

$$d(\alpha) = 100 \times t(\alpha) \times \mathrm{SED}/\mu \quad (10.4)$$

where $t(\alpha)$ is a value from the Normal distribution that is exceeded with probability α, and μ is the mean yield of the trials.

Figure 10.1 presents estimates of critical differences when $\alpha = 0.025$ for different numbers of years and locations, based on mean variance components for European cereal trials (Table 10.1). The critical difference for two years of trials at three locations is 10.5%. Increasing the number of locations from seven to eleven affects the critical differences less than does a change from three to five locations. Generally, gains in precision from adding more trials diminish as the total number of trials increases.

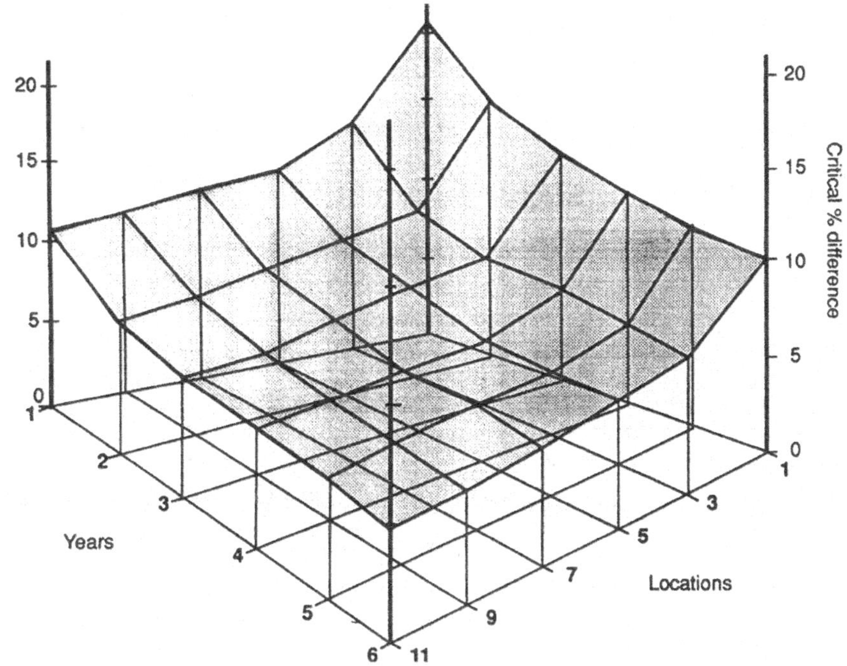

Fig. 10.1. Effect on critical difference ($\alpha = 0.025$) of varying numbers of years and locations (data from official performance testing in European cereal trials).

As demonstrated earlier, for an equivalent change in the total number of trials, a change in the number of years that a genotype is in trial has more influence on precision than changes in the number of locations. For example, three years of trials at three locations gives greater precision than one year of trials at nine locations. Usually, to achieve the greatest precision for a given number of trials, balance in the number of years and locations is necessary. In practice, however, the number of years is limited and gains in precision can come only through increasing the number of locations.

The results in Fig. 10.1 are for trials with two replicates. However, a change in replication has a relatively small effect on precision: an increase in the number of plots per genotype from two to three reduces the critical differences by no more than 0.5%.

The critical percentage differences in Fig. 10.1 are for comparisons of a test genotype with a single check ($c = 1$). An effective way of increasing precision is to compare test genotypes with the mean of several checks. For values of c greater than one the critical differences in Fig. 10.1 are multiplied by a factor as follows.

No. of checks, c	2	3	4	5	6	7
Multiplying factor	0.866	0.816	0.791	0.775	0.764	0.756

10.4 CRITERIA FOR ASSESSING THE EFFICIENCY OF THE PROGRAMME

10.4.1 Matching criteria to objectives

To monitor or tune a system for optimum performance, one needs a gauge to measure progress. Plant breeders often measure their success in terms of sales of seed and numbers of genotypes that are adopted into agricultural practice. Testing authorities will sometimes measure their contribution through changes in national yields per hectare over time, though it is difficult to separate changes in genotype effects from those of GE. However, both these measures of efficiency are too long-term, and broad in scope, to be of immediate help in directing the allocation of trial resources. While critical differences serve as useful measures of precision they take no account of the decision procedures used in promoting genotypes. Here we describe two criteria which combine both precision and information about the rules for selecting genotypes.

Efficiency criteria should reflect the objectives of the system they are intended to monitor. In selection trials two broad objectives are (i) to select genotypes which will improve performance to the greatest extent and (ii) to maximize the probability that a good genotype will be accepted.

Plant breeders accept that, as a consequence of GE and experimental error it is impossible to identify the truly best genotypes. Incorrect choices are inevitable and good material will be discarded. The best one can hope for is to select a subset of genotypes which represent an improvement over most established genotypes. At the same time it is important for the credibility of the selection process that the likelihood is high that a good genotype will be accepted. The role of statistics is to provide criteria which match the goals of the selection system with the realities of the underlying biological variation. Two statistical criteria which might be used for this purpose are potential gain and acceptance probabilities.

Potential gain measures the average difference in performance among genotypes entering trials and those finally selected (Fig. 10.2). Gain is a function of the proportion of genotypes accepted as well as of the accuracy of the trials (see Section 1.3). For a fixed proportion of genotypes accepted, the gain is larger from more accurate trials.

Numerical integration methods (Robinson, 1984) allow us to obtain estimates of the average potential gain to be achieved in trials systems with varying levels of precision and selection rates. The advantages of selective

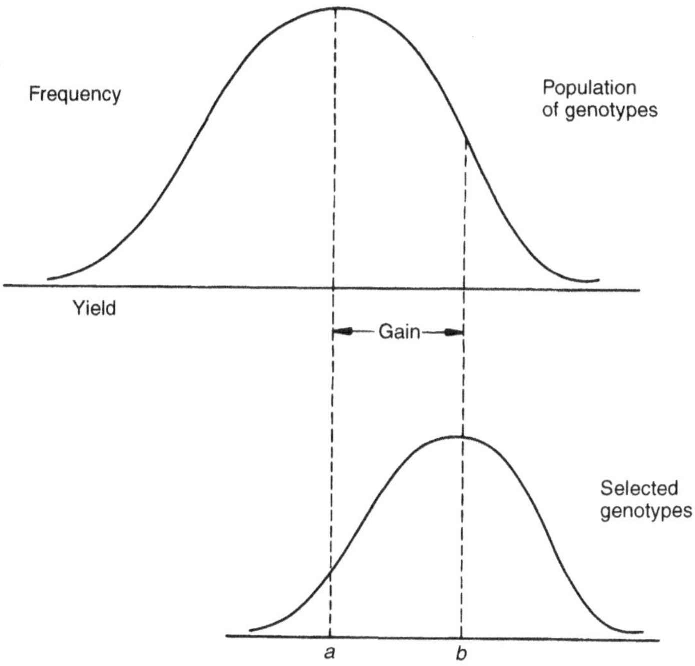

Fig. 10.2. Potential gain from the selection expressed as the % expected difference in mean yield of selected genotypes relative to initial populations, i.e. 100 [$(b - a)/a$].

discarding of genotypes may be seen in Table 10.3. After the first year of testing as many as 60% of the poorer genotypes can be removed without a noticeable reduction in average gain. As a consequence, considerable savings could be made in the total number of plots sown. While there are advantages in selectively discarding genotypes it is important not to discard too many. There is a noticeable reduction in percentage average gain from rejecting as many as 90% of the genotypes after one year of trials. A rough guide to optimal selection is to discard the same proportion at every stage (Finney, 1966).

An acceptance probability is the probability that a genotype of known performance relative to the standard will be accepted. Thus, the probability is influenced by the accuracy of the trials as well as, for any genotype, the size of the difference between its true performance and the acceptance standard (Fig. 10.3).

It can be a salutary lesson to find how easily a good genotype can be missed. Using the averages of the components of variance in Table 10.1 as an example, one can simulate (Robinson, 1984) the risk of a good genotype being rejected. For simplicity, assume we are dealing with a population of genotypes whose yields are Normally distributed and range 5% either side of the population mean and that there are two years of trials with all genotypes being sown in each trial.

Table 10.4 gives the risk of a good genotype being classed as poor or

Table 10.3. Effect of changes in discarding rate and trial precision on potential gain

Year		Total % genotypes discarded by end of year						
1		90	–	60	80	–	60	80
2			90	90	90	–	80	–
3						90	90	90
% Range of genotype difference[a]	*% SE of genotype difference ÷ √2*	*% Average potential gain*						
5	1	2.4	2.7	2.7	2.7	2.8	2.8	2.8
	2	1.6	2.1	2.0	2.0	2.3	2.2	2.1
	3	1.2	1.6	1.6	1.5	1.8	1.8	1.6
	4	0.9	1.2	1.2	1.2	1.5	1.4	1.3
	5	0.7	1.0	1.0	1.0	1.2	1.2	1.1
10	1	4.4	4.5	4.5	4.5	4.6	4.6	4.6
	2	3.6	4.1	4.1	4.0	4.2	4.2	4.2
	3	3.0	3.6	3.5	3.4	3.8	3.8	3.6
	4	2.4	3.1	3.1	2.9	3.4	3.4	3.2
	5	2.1	2.7	2.7	2.5	3.0	3.0	2.8

[a] Range in the performance of genotypes from which selection is being made

mediocre; i.e. being ranked in trials in the lowest one-third or two-thirds of genotypes in trial. A good genotype is defined as having a true yield in the top 1, 5 or 10%. The precision of assessment is defined in terms of % SED or critical difference (Equation 10.4). As the precision of the trial system is reduced, so the probability of rejection increases. Trials at ten locations over two years produce critical differences of 7%. The chance that the fifth best genotype out of 100 will be ranked in trials below the top third is 30%. The risk that it will be ranked in the lowest third is 7%. A corollary is that, when trials become less accurate, there is an increased risk of wrongly selecting poor genotypes.

Acceptance probabilities illustrate more clearly than critical differences the consequences of changing the number of trials. For example, a reduction in the number of locations from twelve to three in a two-year selection programme increases the critical differences by about a half, but more than doubles the risk of a good genotype failing to achieve selection standards.

Critical differences, potential gain and acceptance probabilities are closely related. Critical differences measure the precision of the trials but take no account of the decision procedures, e.g. minimum standards, for promoting genotypes. Acceptance probabilities incorporate a decision rule that leads to acceptance of all genotypes above a standard. Potential gain uses a decision rule that accepts a fixed proportion of the best genotypes. The three criteria are primarily measures of precision and the relative costs and benefits of trials are not part of the criteria, although the benefits can be deduced from the acceptance probabilities and percentage gain. The potential gain indicates what would happen if all of the selected genotypes are grown to the same

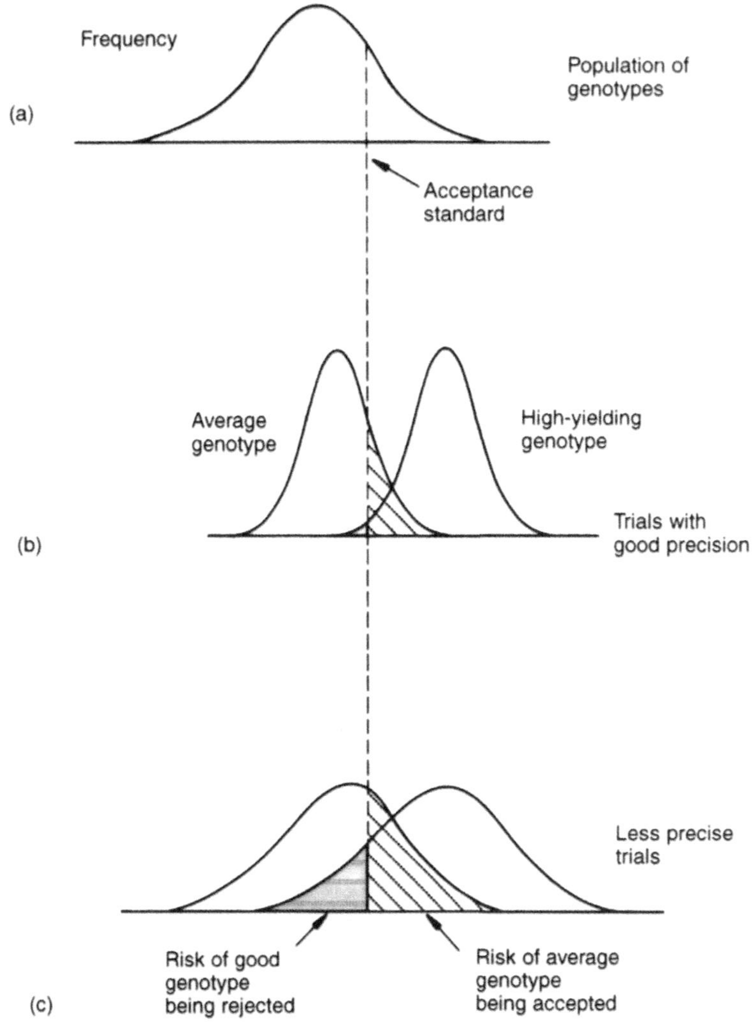

Fig. 10.3. Risks of omitting a good genotype and selecting a mediocre genotype when selection is based on a specified acceptance standard: (a) is the distribution of expected yields for all genotypes; (b) is the expected distribution of trial mean yields for two genotypes from trials with good precision; (c) is the equivalent distribution for less precise trials.

extent. In practice, the best genotypes are more widely accepted in agriculture, and more likely to be selected whatever trials system operates.

10.4.2 Cost implications for selection trials

Several options are available for changing a selection programme. Where the number of locations is reasonably large, i.e. more than five, a small change in

Table 10.4. Risk of a good genotype producing yields in trials which rank it in the lowest third or two-thirds of genotypes after two years of trials

% Range of genotype differences[a]	% SE of genotype difference ÷ √2	True rank of genotype (out of 100)					
		1	5	10	1	5	10
		% Risk of genotype being ranked in lowest two-thirds			% Risk of genotype being ranked in lowest third		
5	1	0	1	3	0	0	0
	2	5	12	19	0	1	2
	3	16	25	31	2	5	7
	4	25	34	39	5	9	11
	5	32	40	44	8	12	14
10	1	0	0	0	0	0	0
	2	0	1	3	0	0	0
	3	1	6	11	0	0	1
	4	5	12	19	0	1	2
	5	10	19	26	1	3	4

[a] Range in the performance of genotypes from which selection is being made.

the number of locations will have limited effect on the precision of the trials system while changing the total cost of the programme broadly in proportion to the number of locations involved.

A change in the testing period will produce a significant change in the precision. However, total testing costs are unlikely to be changed substantially, since, if the genotypes from each stage are sown in the one trial, it is the size of each trial rather than the number of trials that will be affected and the fixed costs of running a trial at a location represent a high proportion of total costs. Of course, there may be substantial costs if the release of a good new variety is delayed. A delay in the case of an official testing programme will affect all varieties equally, and so the cost is to society which cannot rapidly get the benefits of the best material. A delay in an individual breeding programme could mean that varieties produced by other teams continually take the lead in the market place.

A change in the number of replicates is unlikely to affect the precision greatly unless trials are reduced to a single replicate, in which case it becomes difficult to separate the effects of adaptability of genotypes to conditions at the location from plot-to-plot variation (Chapter 6).

In evaluating a trials system, the gain and acceptance probabilities are not alternatives; each describes separate aspects of the same trials system. In the case of official selection trials, acceptance probabilities deal with risks to the breeder while potential gain is concerned with benefits to society. Both aspects are important and should be considered.

The criteria, critical difference and acceptance probability, have useful roles in official testing programmes where the concept of fairness is important. A critical difference can establish minimum performance standards. Acceptance probabilities can help in achieving a balance between the national interest in avoiding the selection of a mediocre genotype and the individual breeder's wishes to minimize the risk of a discarding a good genotype.

10.5 OTHER ISSUES IN RESOURCE ALLOCATION

10.5.1 Genotype selection for specific conditions

The goal of many breeding programmes is to obtain genotypes which are adapted to a wide range of environments. However, can such programmes also serve to identify adaptation for specific conditions, e.g. regional adaptation, tolerance of stress, suitability for soil types, and, if so, how are the resources best distributed amongst locations for this purpose?

Where adaptation is sought in respect of an environmental variable it is sensible to distribute trials so that they provide a representative sample of the locations for the variable in question. The response of genotypes can then be measured by regressing their yields on the environmental variable.

If adaptation is being sought in respect of a categorical environmental variable, e.g. regions, then one needs to test at sufficient locations in each region to provide reliable information for detecting regional differences. In planning for this it is desirable to decide the size of the response which is to be identified. One can then use the components of variation with the formula in Equation 10.3 to estimate the minimum number of trials required to detect a statistically significant interaction.

For example, using the average components of variation in Table 10.1, it is possible to obtain estimates of the relative percentage of regional difference in genotype performance detectable in a two-year trials programme with five check varieties, two replicates per trial, and assuming 10 trial locations in one region and n in the other, e.g.

No. of locations in region (n)	1	2	5	10
Critical diff. % ($P = 0.05$)	9	7	6	5

As a rough rule-of-thumb, to achieve regional recommendations of the same quality as wider recommendations requires almost as many locations within regions as are needed overall.

10.5.2 Compound measurements

For many agricultural crops, the major measure of performance is a compound of gross yield, e.g. fresh weight, and an analytical measure, e.g. dry matter content. The measurement of gross yield is usually straightforward but the analytical measure requires special treatment, for example, a sample must be taken to a laboratory for analysis. The variance of the analytical measure is usually small relative to that of the gross measure and some savings can be made by not measuring the analytical measure on every replicate plot. However, unless the cost of the analytical measure composing the main performance character is substantial then it is wise to sample at least two replicates. In the case of subsidiary analytical measures, e.g. digestible organic matter in forage crops, one analysis of a pooled sample from the replicates is likely to be adequate.

In a multilocation trial involving a costly analytical measure, an alternative form of economy is worth considering. This involves taking samples from a subset of the locations and estimating the analytical measures for the unsampled locations from those for the sampled locations. Finney and Nissen (1956) provide formulae and tables which estimate the most economic ratios of sampled to unsampled locations. Their tables are based on the assumption that the sampled locations are selected at random and adherence to this assumption is often impractical. Nevertheless, sensibly applied, such incomplete recording can contribute to the cost effectiveness of a trials programme.

10.5.3 Useful references

Yates (1940), Cochran (1951), Finney (1958, 1966), Curnow (1961) and Federer (1963) pioneered work on statistical aspects of selection systems. Young (1972, 1976) gave formulae for estimating the potential gain when using information from trials over two seasons. These methods were implemented in a computer program by Robinson and Alonso Sanz (1989).

Components of GE variation have been used in numerous studies to estimate the accuracy of genotype means for trial series with a range of numbers of years and locations. Useful references include Sprague and Federer (1951) and Patterson *et al.* (1977).

10.6 CONCLUSIONS

Plant breeding costs time and effort and all plant breeders strive for cost effectiveness in the allocation of resources. The key issue is how to reduce a very large number of genotypes to the best few as rapidly and efficiently as possible. Broad recommendations are as follows.

- Devote approximately the same resources to each stage of the selection, reducing the number of genotypes at each stage while widening the range of environments in which the survivors are tested.
- Given the small number of years available for testing, and the dominant effect of the GY component of variation, there is little point in an extensive series of trials with a high proportion of genotypes retained throughout.
- Increasing the number of replicates per location beyond three is almost always less beneficial in terms of precision than equivalent increases in the number of locations or years.

References

Anderson, J.R. (1974) Risk efficiency in the interpretation of agricultural production research. *Review of Marketing Agricultural Economics*, **42**, 131–84.

Atlin, G.N. and Frey, K.J. (1989) Breeding crop varieties for low-input agriculture. *American Journal of Alternative Agriculture*, **4**, 53–8.

Austin, R.B., Ford, M.A. and Morgan, C.L. (1989) Genetic improvement in the yield of winter wheat: a further evaluation. *Journal of Agricultural Science, Cambridge*, **112**, 295–301.

Baker, R.J. (1988) Tests for crossover genotype–environmental interactions. *Canadian Journal of Plant Science*, **68**, 405–10.

Basford, K., Kroonenburg, P.M., DeLacy, I.H. and Lawrence, P.K. (1990) Multiattribute evaluation of regional cotton trials. *Theoretical and Applied Genetics*, **79**, 225–34.

Batchelor, L.D. and Reed, H.S. (1918) Relation of the variability of yields of fruit trees to the accuracy of field trials. *Journal of Agricultural Research*, **12**, 245–83.

Baum, W.C. (1986) *Partners Against Hunger*, Consultative Group on International Agricultural Research. World Bank, Washington.

Becker, H.C. (1981) Correlations among some statistical measures of phenotypic stability. *Euphytica*, **30**, 835–40.

Besag, J.E. and Kempton, R.A. (1986) Statistical analysis of field experiments using neighbouring plots. *Biometrics*, **42**, 231–51.

Bingham, J., Blackman, J.A., Angus, W.J. and Longbottom, W. (1981) Breeding and management of wheat varieties, in *The Yield of Cereals 1981*, National Agricultural Centre Cereal Unit, Stoneleigh, UK, pp. 16–30.

Binswanger, H.P. and Barah, B.C. (1980) Yield risk, risk aversion, and genotype selection: conceptual issues and approaches. *Research Bulletin No. 3*. ICRISAT, Patancheru, India.

Bolaños, J. and Edmeades, G.O. (1993) Eight cycles of selection for drought tolerance in lowland tropical maize. II. Responses in reproductive behavior. *Field Crops Research*, **31**, 253–68.

Bos, I. (1983a) The optimum number of replications when testing lines or families on a fixed number of plots. *Euphytica*, **32**, 311–8.

Bos, I. (1983b) About the efficiency of grid selection. *Euphytica*, **32**, 885–93.

Bosemark, N.O. (1993) The need for a comprehensive plant breeding strategy, in *Plant Breeding: Principles and Prospects* (eds M.D. Hayward, N.O. Bosemark and I. Romagosa), Chapman & Hall, London, pp. 525–33.

Bowen, K.L., Teng, P.S. and Roelf, A.P. (1984) Negative interplot interference in field experiments with leaf rust of wheat. *Phytopathology*, **74**, 1157–61.

Box, G.E.P. and Jenkins, G.M. (1970) *Time Series Analysis, Forecasting and Control*, Holden Day, San Francisco.

Bradshaw, J.E. (1986) Competition between cultivars of fodder kale in yield trials with single-row plots. *Euphytica*, **35**, 433–9.

Bradshaw, J.E. (1989) Interplot competition in yield trials of swedes. *Euphytica*, **42**, 135–40.

Braun, H.-J., Pfeiffer, W.H. and Pollmer, W.G. (1992) Environments for selecting widely adapted spring wheat. *Crop Science*, **32**, 1420–7.

Brown, H.K. and Kempton, R.A. (1994) The application of REML in clinical trials. *Statistics in Medicine*, **13**, 1601–17.

Byth, D.E., Eisemann, R.L. and DeLacy, I.H. (1976) Two-way pattern analysis of a large data set to evaluate genotypic adaptation. *Heredity*, **37**, 215–30.

Campbell, L.G. and Lafever, H.N. (1977) Cultivar × environment interaction in soft red winter wheat yield tests. *Crop Science*, **17**, 604–8.

Ceccarelli, S. (1994) Specific adaptation and breeding for marginal conditions. *Euphytica*, **77**, 205–19.

Christidis, B.G. (1939) Further studies on competition in yield-trials with cotton. *Empire Journal of Experimental Agriculture*, **7**, 111–20.

CIMMYT (1993) *1992/93 CIMMYT World Wheat Facts and Trends. The Wheat Breeding Industry in Developing Countries: An Analysis of Investments and Impacts*, CIMMYT, Singapore.

Cleaver, T.J., Greenwood, D.J. and Wood, J.T. (1970) Systematically arranged fertilizer experiments. *Journal of Horticultural Science*, **45**, 457–69.

Cochran, W.G. (1951) Improvement by means of selection, in *Proceedings of the 2nd Berkeley Symposium on Mathematics, Statistics and Probability* (ed. J. Neyman), University of California Press, pp. 449–70.

Connolly, T., Currie, I.D., Bradshaw, J.E. and McNicol, J.W. (1993) Interplot competition in yield trials of potatoes (*Solanum tuberosum* L.) with single-drill plots. *Annals of Applied Biology*, **123**, 367–77.

Cooper, M. and DeLacy, I.H. (1994) Relationships among analytical methods used to study genotypic variation and genotype-by-environment interaction in plant breeding multi-environment experiments. *Theoretical and Applied Genetics*, **88**, 561–72.

Cooper, M., DeLacy, I.H., Byth, D.E. and Woodruff, D.R. (1993) Predicting grain yield in Australian environments using data from CIMMYT international wheat performance trials. 2. The application for classification to identify environmental relationships which exploit correlated response to selection. *Field Crops Research*, **32**, 323–42.

Cressie, N. (1988) A graphical procedure for determining nonstationarity in time series. *Journal of the American Statistical Association*, **83**, 1108–16.

Crossa, J. (1990) Statistical analyses of multilocation trials. *Advances in Agronomy*, **44**, 55–85.

Crossa, J., Basford, K., Taba, S. *et al.* (1995) Three-mode analyses of maize using morphological and agronomic attributes measured in multilocational trials. *Crop Science*, **35**, 1483–91.

Crossa, J., Fox, P.N., Pfeiffer, W.H. *et al.* (1991) AMMI adjustment for statistical analysis of an international wheat yield trial. *Theoretical and Applied Genetics*, **81**, 27–37.

Cullis, B.R. and Gleeson, A.C. (1991) Spatial analysis of field experiments – an extension to two dimensions. *Biometrics*, **47**, 1149–460.

Cullis, B.R., Gleeson, A.C. and Thomson, F.M. (1992) The response to selection of different procedures for the analysis of early generation variety trials. *Journal of Agricultural Science, Cambridge*, **118**, 141–8.

Cullis, B.R., Lill, W.J., Fisher, J.A. *et al.* (1989) A new procedure for the analysis of early generation variety trials. *Applied Statistics*, **38**, 361–75.

Cullis, B.R., McGilchrist, C.A. and Gleeson, A.C. (1991) Error model diagnostics in the general linear model with applications to the analysis of repeated measurements and field experiments. *Journal of the Royal Statistical Society B*, **53**, 409–16.

Curnow, R.N. (1961) Optimal programmes for varietal selection. *Journal of the Royal Statistical Society B*, **23**, 282–318.

David, O. and Kempton, R.A. (1996) Designs for interference. *Biometrics*, **52**, 224–233.

David, O., Kempton, R.A. and Nevison, I.M. (1996) Designs to control interplot competition in variety trials. *Journal of Agricultural Science, Cambridge*, **127**, 285–8.

DeLacy, I.H., Eisemann, R.L. and Cooper, M. (1990) The importance of genotype-by-environment interaction in regional genotype trials, in *Genotype-By-Environment Interaction and Plant Breeding* (ed. M.S. Kang), Louisiana State University, Baton Rouge, pp. 287–300.

DeLacy, I.H., Fox, P.N., Corbett, J.D. *et al.* (1994) Long-term association of locations for testing spring bread wheat. *Euphytica*, **72**, 95–106.

Dickerson, G.E. (1962) Implications of genetic–environmental interaction in animal breeding. *Animal Production*, **4**, 47–63.

Dickson, J.M. and Talbot, M. (1986) Statistical data validation and expert systems, in *Proceedings of COMPSTAT Symposium*, Rome, Physica-Verlag, Heidelberg, pp. 283–8.

Digby, P.G.N. (1979) Modified joint regression analysis for incomplete variety × environment data. *Journal of Agricultural Science, Cambridge*, **93**, 81–6.

Digby, P.G.N. and Kempton, R.A. (1987) *Multivariate Analysis of Ecological Communities*, Chapman & Hall, London.

Dyke, G.V. and Shelley, C.F. (1976) Serial designs balanced for effects of neighbours on both sides. *Journal of Agricultural Science, Cambridge*, **87**, 303–5.

Eskridge, K.M. (1990) Selection of stable cultivars using safety-first rule. *Crop Science*, **30**, 369–74.

Eskridge, K.M., Byrne, P.F. and Crossa, J. (1991) Selection of stable varieties by minimizing the probability of disaster. *Field Crop Research*, **27**, 169–81.

Eskridge, K.M. and Mum, R.F. (1992) Choosing cultivars based on the probability of outperforming a check. *Theoretical and Applied Genetics*, **84**, 494–500.

Falconer, D.S. (1981) *Introduction to Quantitative Genetics*, 2nd edn, Longman Scientific & Technical, London.

Fasoulas, A. (1973) *A New Approach to Breeding Superior Yielding Varieties.* Aristotelian University of Thessaloniki, Department of Genetics and Plant Breeding, publication 3, Thessaloniki, Greece.

Fasoulas, D.A. (1990) Correlations between auto-, allo- and nil-competition and their implications in plant breeding. *Euphytica*, **50**, 57–62.

Federer, W.T. (1961) Augmented designs with one-way elimination of heterogeneity. *Biometrics*, **17**, 447–73.

Federer, W.T. (1963) Procedures and designs useful for screening material in selection and allocation, with a bibliography. *Biometrics*, **19**, 553–87.

Federer, W.T. and Raghavarao, D. (1975) On augmented designs. *Biometrics*, **31**, 29–35.

Finlay, K.W. and Wilkinson, G.N. (1963) The analysis of adaptation in a plant-breeding programme. *Australian Journal of Agricultural Research*, **14**, 742–54.

Finney, D.J. (1958) Statistical problems of plant selection. *Bulletin of the Institute of International Statistics*, **36**, 242–68.

Finney, D.J. (1966) An experimental study of certain screening processes. *Journal of the Royal Statistical Society B*, **28**, 88–109.

Finney, D.J. and Nissen, O. (1956) Incomplete determination of a measure of quality in a series of experiments. *Journal of Agricultural Science, Cambridge*, **48**, 124–8.

Fischer, R.A. (1979) Are your results confounded by intergenotypic competition?, in *Proceedings of Vth International Wheat Genetics Symposium*, Vol. 2, (ed. S. Ramanujam), Indian Society of Genetics and Plant Breeding, New Delhi, pp. 767–77.

Fisher, R.A. (1935) *The Design of Experiments*, Oliver and Boyd, Edinburgh.

Fox, P.N. and Rosielle, A.A. (1982a) Reducing the influence of environmental main-effects on pattern analysis of plant breeding environments. *Euphytica*, **31**, 645–56.

Fox, P.N. and Rosielle, A.A. (1982b) Reference sets of genotypes and selection for yield in unpredictable environments. *Crop Science*, **22**, 1171–75.

Fox, P.N., Rosielle, A.A. and Boyd, W.J.R. (1985) The nature of genotype × environment interactions for wheat yield in Western Australia. *Field Crops Research*, **11**, 387–98.

Fox, P.N., Skovmand, B., Thompson, B.K. *et al.* (1990) Yield and adaptation of hexaploid spring triticale. *Euphytica*, **47**, 57–64.

Freeman, G.H. (1973) Statistical methods for the analysis of genotype–environment interactions. *Heredity*, **31**, 339–54.

Freeman, G.H. (1979) Some two-dimensional designs balanced for nearest neighbours. *Journal of the Royal Statistical Society B*, **41**, 88–95.

Gardiner, C.O. (1961) An evaluation of effects of mass selection and seed irradiation with thermal neutrons on yield of corn. *Crop Science*, **1**, 241–5.

Gauch, H.G. (1988) Model selection and validation for yield trials with interaction. *Biometrics*, **44**, 705–15.

Gauch, H.G. (1992) *Statistical Analysis of Regional Yield Trials: AMMI Analysis of Factorial Designs*, Elsevier, Amsterdam.

Gauch, H.G. and Zobel, R.W. (1988) Predictive and postdictive success of statistical analyses of yield trials. *Theoretical and Applied Genetics*, **76**, 1–10.

Gilmour, A. (1992) *TwoD: a Program to Fit a Mixed Linear Model with Two Dimensional Spatial Adjustment for Local Trend*. NSW Agriculture, Orange NSW 2800, Australia.

Gleeson, A.C. and Cullis, B.R. (1987) Residual maximum likelihood (REML) estimation of a neighbour model for field experiments. *Biometrics*, **43**, 277–88.

Gleeson, A.C. and Eccleston, J.A. (1992) On the design of field experiments under a one-dimensional neighbour model. *Australian Journal of Statistics*, **34**, 91–7.

Gollob, H.F. (1968) A statistical test model which combines features of factor analytic and analysis of variance techniques. *Psychometrika*, **33**, 72–115.

Gomez, K.A. (1972) Border effects in rice experimental plots. II. Varietal competition. *Experimental Agriculture*, **8**, 295–8.

Green, P.J., Jennison, C. and Seheult, A.H. (1985) Analysis of field experiments by least squares smoothing. *Journal of the Royal Statistical Society B*, **47**, 299–315.

Healey, R.G. (1991) Database management systems, in *Geographical Information Systems, Vol. 1, Principles* (eds D.J. Maguire, M.F. Goodchild and D.W. Rhind), Longman Scientific and Technical Ltd., New York, pp. 251–67.

Hemmerle, W.J. and Downs, B.W. (1978) Non-homogeneous variances in the mixed AOV model; maximum likelihood estimation, in *Contributions to Survey Sampling and Applied Statistics* (ed. H.A. David), Academic Press, New York.

Holtsmark, G. and Larsen, B.R. (1905) Om muligheder for at indskraenke de fejl, som ved marksforsog betinges af jordens uensartethed. *Tidsskrift for Landbrugets Planteavl*, **12**, 330–51.

Immer, F.R., Hayes, H.K. and Powers, Le Roy (1934) Statistical determination of barley variety adaptation. *Journal of American Society of Agronomy*, **26**, 403–19.

Jenkyn, J.F. and Bainbridge, A. (1974) Disease gradients and small plot experiments on barley mildew. *Annals of Applied Biology*, **76**, 269–79.

Jensen, N.F. and Federer, W.T. (1964) Adjacent row competition in wheat. *Crop Science*, **4**, 641–5.

Kaufmann, M.L. and McFadden, A.D. (1960) The competitive interaction between barley plants grown from large and small seeds. *Canadian Journal of Plant Science*, **40**, 623–9.

Kawano, K., Amaya, A., Daza, P. and Rios, M. (1978) Factors affecting efficiency of hybridization and selection in cassava. *Crop Science*, **18**, 373–6.

Kempton, R.A. (1982) Adjustment for competition between varieties in plant breeding trials. *Journal of Agricultural Science, Cambridge*, **98**, 599–611.

Kempton, R.A. (1984a) The design and analysis of unreplicated trials. *Vortage fur Pflanzenzüchtung*, **7**, 219–42.

Kempton, R.A. (1984b) The use of biplots in interpreting variety by environment interactions. *Journal of Agricultural Science, Cambridge*, **103**, 123–35.

Kempton, R.A. (1985) Statistical models for interplot competition. *Aspects of Applied Biology*, **10**, 111–20.

Kempton, R.A. (1992) Interference in agricultural experiments, in *Proceedings of the 2nd Meeting of the Biometric Society East, Central & Southern African Network* (eds R. Kempton, R. Coe and Sr. J. Canhão), International Biometric Society, Harare, Zimbabwe, pp. 1–16.

Kempton, R.A. (1994) Review of 'Statistical Analysis of Regional Yield Trials' by H.G. Gauch. *Heredity*, **73**, 220–2.

Kempton, R.A., Gregory, R.S., Hughes, W.G. and Stoehr, P.J. (1986) The effect of interplot competition on yield assessment in triticale trials. *Euphytica*, **35**, 257–65.

Kempton, R.A. and Lockwood, G. (1984) Interplot competition in variety trials of field beans (*Vicia faba* L.). *Journal of Agricultural Science, Cambridge*, **103**, 293–301.

Kempton, R.A., Seraphin, J.C. and Sword, A.M. (1994) Statistical analysis of two-dimensional variation in variety yield trials. *Journal of Agricultural Science, Cambridge*, **122**, 335–42.

Kempton, R.A. and Talbot, M. (1988) The development of new crop varieties. *Journal of the Royal Statistical Society A*, **151**, 327–41.

Knight, R. (1970) The measurement and interpretation of genotype–environment interactions. *Euphytica*, **19**, 225–35.

Kramer, Th., van Ooijen, J.W. and Spitters, C.J.T. (1982) Selection for yield in small plots of spring wheat. *Euphytica*, **31**, 549–64.

Kroonenberg, P.M. and Basford, K. (1989) An investigation of multi-attribute genotype response across environments using three-mode principal component analysis. *Euphytica*, **44**, 109–23.

Lill, W.J., Gleeson, A.C. and Cullis, B.R. (1988) Relative accuracy of a neighbour model for field experiments. *Journal of Agricultural Science, Cambridge*, **111**, 339–46.

Lin, C.S. and Binns, M.R. (1985) Procedural approach for assessing cultivar × location data. Pairwise genotype–environment interactions of test cultivars with checks. *Canadian Journal of Plant Science*, **65**, 1065–71.

Lin, C.S. and Poushinsky, G. (1983) A modified augmented design for an early stage of plant selection involving a large number of test lines without replication. *Biometrics*, **39**, 553–61.

Lin, C.S., Poushinsky, G. and Voldeng, H.D. (1985) Design and model for investigating competition effects from neighbouring test plots. *Canadian Journal of Plant Science*, **65**, 1073–7.

Lipps, P.E. and Madden, L.V. (1992) Effects of plot size and border width on assessment of powdery mildew of winter wheat. *Plant Disease*, **76**, 299–303.

Ljung, G.M. (1986) Diagnostic testing of univariate time series models. *Biometrika*, **73**, 725–30.

Martin, R.J. (1990) The use of time-series models and methods in the analysis of agricultural field trials. *Communications of Statistical Theory and Methods*, **19**, 55–81.

McCullagh, P. and Nelder, J.A. (1983) *Generalised Linear Models*, Chapman & Hall, London.

McGilchrist, C.A. (1965) Analysis of competition experiments. *Biometrics*, **21**, 975–85.

Mead, R. (1988) *The Design of Experiments: Statistical Principles for Practical Applications*, Cambridge University Press, Cambridge.

Mead, R., Curnow, R.N. and Hasted, A. (1993) *Statistical Methods in Agriculture and Experimental Biology*, Chapman & Hall, London.

Menz, K.M. (1980) A comparative analysis of wheat adaptation across international environments using stochastic dominance and pattern analysis. *Field Crops Research*, **3**, 33–41.

Nelder, J.A. (1965) The analysis of randomised experiments with orthogonal block structure. I. Block structure and the null analysis of variance. *Proceedings of the Royal Society A*, **283**, 147–62.

Nguyen, N.K. and Williams, E.R. (1993) An algorithm for constructing resolvable row-column designs with $v \leq 100$. *Australian Journal of Statistics*, **35**, 360–70.

Parlevliet, J.E. and van Ommeren, A. (1984) Interplot interference and the assessment of barley cultivars for partial resistance of leaf rust, *Puccinia hordei*. *Euphytica*, **33**, 685–97.

Patterson, H.D. (1978) Routine least squares estimation of variety means in incomplete tables. *Journal of the National Institute of Agricultural Botany*, **13**, 142–51.

Patterson, H.D. (1982) FITCON and the analysis of incomplete varieties × trials tables. *Utilitas Mathematica*, **21A**, 267–89.

Patterson, H.D. and Hunter, E.A. (1983) The efficiency of incomplete block designs in National List and Recommended List cereal variety trials. *Journal of Agricultural Science, Cambridge*, **101**, 427–33.

Patterson, H.D. and Nabugoomu, F. (1992) REML and the analysis of series of crop variety trials. *Proceedings of the XVIth International Biometric Conference*, Hamilton, NZ, pp. 77–93.

Patterson, H.D. and Silvey, V. (1980) Statutory and recommended list trials of crop varieties in the United Kingdom. *Journal of the Royal Statistical Society A*, **143**, 219–52.

Patterson, H.D, Silvey, V., Talbot, M. and Weatherup, S.T.C. (1977) Variability of yields of cereal varieties in UK trials. *Journal of Agricultural Science, Cambridge*, **89**, 239–45.

Patterson, H.D. and Thompson, R. (1971) Recovery of inter-block information when block sizes are unequal. *Biometrika*, **58**, 545–54.

Patterson, H.D. and Thompson, R. (1974) Maximum likelihood estimation of components of variance, in *Proceedings of the 8th International Biometric Conference*, Costanza, pp. 197–207.

Patterson, H.D., Thompson, R., Hunter, E.A. and Kempton, R.A. (1990) *Analysis of Unbalanced Data Using REML*, Scottish Agricultural Statistics Service, Edinburgh.

Patterson, H.D. and Williams, E.R. (1976) A new class of resolvable incomplete block designs. *Biometrika*, **63**, 83–90.

Patterson, H.D., Williams, E.R. and Hunter, E.A. (1978) Block designs for variety trials. *Journal of Agricultural Science, Cambridge*, **90**, 395–400.

Paysour, R.E. and Fry, W.E. (1983) Interplot interference – a model for planning field experiments with aerially disseminated pathogens. *Phytopathology*, **73**, 1014–20.

Peterson, C.J. and Pfeiffer, W.H. (1989) International winter wheat evaluation: Relationships among test sites based on cultivar performance. *Crop Science*, **29**, 276–82.

Pritchard, F.J. (1916) The use of checks and repeated plantings in varietal tests. *Journal of the American Society of Agronomy*, **8**, 65–81.

Randle, W.M., Davis, D.W. and Groth, J.V. (1986) Interplot interference in field plots with leaf rust of maize. *Journal of American Society of Horticultural Science*, **111**, 297–300.

Rich, P.A. (1973) Influence of cultivar, row spacing and number of rows on yield of wheat plots. *Agronomy Journal*, **65**, 331–3.

Richey, F.D. (1924) Adjusting yields to their regression on a moving average as a means of correcting for soil heterogeneity. *Journal of Agricultural Research*, **27**, 79–90.

Richey, F.D. (1926) The moving average as a basis for measuring correlated variation in agronomic experiments. *Journal of Agricultural Research*, **32**, 1161–75.

Riggs, T.J. (1986) Collaborative spring barley trials in Europe 1980–82. Analysis of grain yield. *Zeitschrift für Pflanzenzüchtung*, **96**, 289–303.

Robinson, D.L. (1984) A study of sequential variety selection systems. *Journal of Agricultural Science, Cambridge*, **102**, 315–21.

Robinson, D.L. and Alonso-Sanz, R. (1989) *Selsys User Manual*, BioSS, Edinburgh.

Robinson, D.L., Kershaw, C.D. and Ellis, P.R. (1988) An investigation of two-dimensional yield variability in breeders' small plot trials. *Journal of Agricultural Science, Cambridge*, **111**, 419–26.

Roemer, T. (1930) *Der Feldversuch*, 3rd edn, DLG-Verlag, Frankfurt a.M.

Romagosa, I. and Fox, P.N. (1993) Genotype × environment interaction and adaptation, in *Plant Breeding: Principles and Prospects* (eds M.D. Hayward, N.O. Bosemark and I. Romagosa), Chapman & Hall, London, pp. 373–90.

Roy, A.D. (1952) Safety-first and the holding of assets. *Econometrica*, **20**, 431–49.

Royo, C., Rodriguez, A. and Romagosa, I. (1993) Differential adaptation of complete and substituted triticale. *Plant Breeding*, **111**, 113–9.

Salmon, S.C. (1914) Check plots as a source of error in varietal tests. *Journal of the American Society of Agronomy*, **6**, 128–31.

Schutz, W.M. and Cockerham, C.C. (1966) The effect of field blocking on gain from selection. *Biometrics*, **22**, 843–63.

Silvey, V. (1978) Methods of analysing NIAB variety trials over many sites and several seasons. *Journal of the National Institute of Agricultural Botany*, **14**, 385–400.

Silvey, V. (1986) The contribution of new varieties to cereal yields in England and Wales between 1947 and 1983. *Journal of the National Institute of Agricultural Botany*, **17**, 155–68.

Simmonds, N.W. (1979) *Principles of Crop Improvement*, Longman, London.

Souza, E., Myers, J.R. and Scully, B.T. (1993) Genotype by environment interaction in crop improvement, in *Crop Improvement for Sustainable Agriculture* (eds M.B. Callaway and C.A. Francis), University of Nebraska Press, Lincoln, pp. 192–233.

Spitters, C.J.T. (1979) Competition and its consequences for selection in barley breeding. *Agriculture Research Report 893*, Centre for Agricultural Publishing and Documentation, Wageningen.

Sprague, G.F. and Federer, W.T. (1951) A comparison of variance components in corn yield trials. II. Error, year × variety, location × variety, and variety components. *Agronomy Journal*, **43**, 535–41.

Stadler, L.J. (1921) Experiments in field plot technic for the preliminary determination of comparative yields in the small grains. *University of Missouri Agricultural Experimental Station, Research Bulletin 49.*

Stam, P. (1984) Estimation of genotypic values without replication in field trials. *Euphytica*, **33**, 841–52.

Student (1923) On testing varieties of cereals. *Biometrika*, **13**, 271–95.

Talbot, M. (1984) Yield variability of crop varieties in the United Kingdom. *Journal of Agricultural Science, Cambridge*, **102**, 315–21.

Talbot, M., Milner, A.D., Nutkins, M.A.E. and Law, J.R. (1995) The effect of interference between plots on yield performance in crop variety trials. *Journal of Agricultural Science, Cambridge*, **124**, 335–42.

Thorne, C.E. (1907) The interpretation of field experiments. *Journal of the American Society of Agronomy*, **1**, 45–55.

Trenbath, B.R. (1978) Models and the interpretation of mixture experiments, in *Plant Relations in Pastures* (ed. J.R. Wilson), CSIRO, Melbourne, pp. 145–62.

Ud-Din, N., Carver, B.F. and Clutter, A.C. (1992) Genetic analysis and selection for wheat yield in drought-stressed and irrigated environments. *Euphytica*, **62**, 89–96.

Ullman, J.D. (1982) *Principles of Database Design*, Computer Science Press, Rockville, Maryland.

Van der Plank, J.E. (1963) *Plant Diseases: Epidemics and Control*, Academic Press, New York.

Waggoner, P.E. (1952) Distribution of potato late blight around an inoculum source. *Phytopathology*, **42**, 323–8.

Ward, J.H. (1963) Hierarchical grouping to optimise an objective function. *Journal of the American Statistical Association*, **58**, 236–44.

Wiancko, A.T. (1914) Use and management of check plots in soil fertility investigations. *Journal of the American Society of Agronomy*, **6**, 122–4.

Wiancko, A.T., Arny, A.C. and Salmon, S.C. (1921) Report of committee on standardization of field experiments. *Journal of the American Society of Agronomy*, **13**, 368–74.

Wilkinson, G.N., Eckert, S.R., Hancock, T.W. and Mayo, O. (1983) Nearest neighbour (NN) analysis of field experiments (with discussion). *Journal of the Royal Statistical Society B*, **45**, 151–211.

Wilkinson, G.N. and Rogers, C.E. (1973) Symbolic description of factorial models for analysis of variance. *Applied Statistics*, **22**, 392–9.

Williams, E.J. (1962) The analysis of competition experiments. *Australian Journal of Biological Sciences*, **15**, 509–25.

Williams, E.R. (1986a) A neighbour model for field experiments. *Biometrika*, **73**, 279–87.

Williams, E.R. (1986b) Row and column designs with contiguous replicates. *Australian Journal of Statistics*, **28**, 154–63.

Williams, E.R. and Matheson, A.C. (1994) *Experimental Design and Analysis for Use in Tree Improvement*, CSIRO, Canberra.

Williams, E.R. and Talbot, M. (1993) *ALPHA+ Experimental Designs for Variety Trials*, CSIRO, Canberra and BioSS, Edinburgh.

Williams, R.M. (1952) Experimental designs for serially correlated observations. *Biometrika*, **39**, 151–67.

Williams, W.T. (1976) *Pattern Analysis in Agricultural Science*, Elsevier Scientific Publishing Company, Amsterdam.

Yates, F. (1933) The principles of orthogonality and confounding in replicated experiments. *Journal of Agricultural Science, Cambridge*, **23**, 108–45.

Yates, F. (1936) A new method of arranging variety trials involving a large number of varieties. *Journal of Agricultural Science, Cambridge*, **26**, 424–55.

Yates, F. (1940) Modern experimental design and its function in plant selection. *Empire Journal of Experimental Agriculture*, **31**, 223–30.

Yates, F. and Cochran, W.G. (1938) The analysis of groups of experiments. *Journal of Agricultural Science, Cambridge*, **28**, 556–80.

Young, J.C. (1972) An investigation of procedures for multiple stage selection for a variate subject to errors of measurement. *Biometrika*, **59**, 323–34.

Young, J.C. (1976) Varietal screening from finite normal populations. *Journal of the American Statistical Association*, **71**, 87–92.

Zadoks, J.C., Chang, T.T. and Konzak, C.F. (1974) A decimal code for the growth stages of cereals. *Weed Research*, **14**, 415–21.

Zimmerman, D.L. and Harville, D.A. (1991) A random field approach to the analysis of field-plot experiments and other spatial experiments. *Biometrics*, **47**, 223–39.

Index

Page numbers appearing in *italics* refer to tables and page numbers appearing in **bold** refer to figures.

GPSR Compliance

The European Union's (EU) General Product Safety Regulation (GPSR) is a set of rules that requires consumer products to be safe and our obligations to ensure this.

If you have any concerns about our products, you can contact us on ProductSafety@springernature.com

In case Publisher is established outside the EU, the EU authorized representative is:

Springer Nature Customer Service Center GmbH
Europaplatz 3
69115 Heidelberg, Germany

Batch number: 10809441

Printed by Printforce, the Netherlands